WRITING FOR LAW ENFORCEMENT

Christopher Thaiss
George Mason University

John E. Hess
Federal Bureau of Investigation, Retired

Allyn and Bacon
Boston • London • Toronto • Sydney • Tokyo • Singapore

Vice President, Humanities: Joseph Opiela
Editorial Assistant: Rebecca Ritchey
Executive Marketing Manager: Lisa Kimball
Editorial–Production Administrator: Donna Simons
Editorial–Production Service: Matrix Productions Inc.
Composition and Prepress Buyer: Linda Cox
Manufacturing Buyer: Suzanne Lareau
Cover Administrator: Jenny Hart
Electronic Composition: Cabot Computer Services

Between the time Website information is gathered and then published, it is not
unusual for some sites to have closed. Also, the transcription of URLs can result in
unintended typographical errors. The publisher would appreciate notification
where these occur so that they may be corrected in subsequent editions.

Library of Congress Cataloging-in-Publication Data

Thaiss, Christopher J.
    Writing for law enforcement / Christopher Thaiss, John Hess.
      p.   cm.
    Includes bibliographical references and index.
    ISBN 0-205-28389-6
    1. Police reports—Handbooks, manuals, etc. 2. Technical writing.
  3. Communication in police administration. I. Hess, John E.
  II. Title.
    HV7936.R53T48   1998
    808'.066363—dc21   98-23572
                                          CIP

Printed in the United States of America
10 9 8 7 6 5 4 3 2   03 02 01 00 99 98

CONTENTS

PREFACE

And paperwork? You'll write enough words in
your lifetime to stock a library. You'll learn
to live with doubt, anxiety, frustration.
—SGT. JOE FRIDAY (CHARACTER), DRAGNET*

THE PROBLEM AND THE SOLUTION

Many law enforcement professionals regard writing as a necessary evil. It's the "dreaded paperwork" that's best done quickly and gotten over with. As people of action, police don't relish the quiet, slow work that good writing demands. Moreover, written reports, if not precisely done, can jump up and bite their writers when these reports are used by clever attorneys to undermine a painstaking investigation. No wonder that many police officers wish that they could avoid writing altogether.

Yet good writing is in key ways the backbone of law enforcement. Without good reports, there is no case. Without full and accurate records, there is no history, no precedent. Without clear, careful memos, there is no communication within the organization; efficiency erodes and morale crumbles.

Writing for Law Enforcement takes seriously the professional's need to perform these tasks well and efficiently. Writing, like all other main elements of police work, demands care, precision, and smart work habits. We intend this brief guide to help all law enforcement professionals build writing as a skill that they can use with confidence and no waste of time or effort.

---

*From "What's a Cop?" script by Preston Wood, compiled in *Words of Dragnet*, http://www.hooked.net/~cbhall/DRAGQUOT.HTM, 1997.

HOW TO USE THIS BOOK

This book is not meant to be read cover to cover, though we don't discourage you from doing so. We expect readers to refer to specific chapters as the need arises. That's why we've included a detailed table of contents and an index of topics.

Writing for Law Enforcement addresses two kinds of readers:

1. *The working professional,* who must conduct and write up interviews, searches, and interrogations; who must compile reports; who must write memos; who may be called on to give occasional talks to professional or community groups.
2. *The student of criminal justice,* who must learn to write those same professional documents and also analyze books and articles; who must write papers for classes; who must take essay exams; who must give oral presentations.

We have kept both kinds of readers in mind throughout our writing of the guide. We think that every chapter offers something for both the writer-on-the-job and the writer-in-school—indeed, we recognize that many people in law enforcement must balance both work and school. Certainly, also, the advanced student of criminal justice who interns with law enforcement organizations is both worker and student.

SUMMARY OF CHAPTERS

Chapter 1, "Writing Techniques to Increase Learning," provides tips on ways writing can be used both to *improve reading, listening, and observation skills* and to make *note taking* simpler and more productive. Chapter 1 will help you become a more *confident, fluent communicator* through regular exercises.

Chapter 2, "The Writing Process," takes you through a four-step, time-tested sequence that can be applied to most writing tasks. Learning this process will (1) help you *overcome anxiety* and writer's block, (2) *save time* through effective time management, (3) *remove much of the guesswork* from writing for readers, and (4) *ensure higher quality products* that meet your readers' needs.

Chapter 3, "Writing Investigative Reports: Conducting Research," begins a sequence of four chapters that simplifies report writing by examining each phase of the report-writing process and offering clear advice. Chapter 3 covers various investigative tasks, particularly the interview, that result in those reports. This chapter expands on the idea of the interview as "a conversation with a purpose"; among other techniques, we explain the roles of writing and listening in preparing for, conducting, verifying, and summarizing interviews.

Chapter 4, "Writing Investigative Reports: Predrafting Considerations," describes ways to use writing to develop, clarify, and organize information from interviews and other investigative techniques in order to make report writing more productive and efficient. Outlining and freewriting are among techniques illustrated.

Chapter 5, "Writing Investigative Reports: Format," shows how experienced case officers compile the pertinent documents into clear reports that address the needs of real readers. Formatting tips, such as cover memos for the report and preambles for interviews, searches, arrests, and interrogations, allow writers to make report assembly logical and straightforward.

Chapter 6, "Writing Investigative Reports: Pitfalls," offers advice on style intended to make reports more effective through maintenance of high professional standards. Policespeak, jargon, prejudices, and misuse of quotes are among the obstacles the chapter teaches writers how to avoid.

Chapter 7, "Writing Memos," stresses the importance of tailoring memos to the reader's perspective rather than the writer's and offers suggestions for doing so. We suggest how techniques popular in sales and marketing can increase the success of several types of memos, from the memo of explanation to the memo of request. Samples of three types of memos are included.

Chapter 8, "Oral Presentations," shows how to maximize your effectiveness as a speaker. It also describes ways that informal writing can help you plan, organize, and "script" talks that your listeners will remember.

Chapter 9, "Taking Essay Exams," specifically addresses the student writer. It offers a sequence of planning and writing strategies to help you use exam time most wisely, avoid exam panic, and write informative, well-organized essays.

Chapter 10, "Finding and Citing Sources for Research," briefly lists print and online sources used in typical law enforcement research tasks. It also shows how to cite sources and compile bibliographies in research documents by using the standards of the American Psychological Association (APA).

ACKNOWLEDGMENTS

Thanks to Joe Opiela of Allyn and Bacon for shaping my idea for a multidisciplinary writing text into a series of focused volumes, of which *Writing for Law Enforcement* is one of the first two numbers. This project has brought me into collaboration with John Hess, who has not only taught me valuable lessons about law enforcement methods and theory, but who has also exemplified cooperation, imagination, and courtesy. I am most thankful for having had this opportunity to work with him. Brian Barker, poet and bibliographer, deserves thanks for the careful and creative research that produced our final chapter. Thanks also to the College of Arts and Sciences, Daniele Struppa, Dean, for supporting Brian's work on this project.

At Allyn and Bacon, thanks also to Rebecca Ritchey and Lisa Kimball for their efforts toward the success of this project. For their constructive reviews of the work-in-progress, John and I thank Dale Mooso, San Antonio College; Thomas Petee, Auburn University; and Dr. Jo Ann M. Scott, Ohio Northern University. As always, no words can adequately express my thanks for Ann Louise, Christopher, Flannery, Jeff, Jimmy, and Irene, who always work for justice, nor for Ann, whose example leads the team.

—Christopher Thaiss

My thanks go to Steve Gladis for encouraging my work on this project, to Chris Thaiss, for putting it all together, and to my wife, Jean, for her patience and support.

—John Hess

1

WRITING TECHNIQUES
TO INCREASE LEARNING

*Columbo reached into his raincoat pocket and
pulled out a steno pad. "I oughta be makin' some
notes, " he said. He patted his jacket pockets.
"Pencil . . . I don't know where my pencils go. Mrs.
Columbo always makes sure every morning before I
leave the house that I've got a nice yellow pencil."*

*"I'm taking notes, Lieutenant," said
[Detective Zimmerman].*

*He glanced at her. "So y'are. I appreciate it.
Y'know, I always wonder about those Sherlock
Holmes TV shows. He never takes any notes.
I can't see how he remembers everything."**

While much of this book explains reports, memos, and other formal types
of writing in law enforcement, this chapter focuses on writing tricks and
tools that professionals in the field or the classroom can use informally
to enhance their critical and creative thinking as well as their recall of
information.

---

*From William Harrington, *Columbo: The Hoffa Connection* (New York: Tom Doherty Associates, 1995), p. 43.

1

A QUESTION OF ATTITUDE:
WRITING FOR YOURSELF

When your goal is to improve thinking and learning, it's basic that you experiment with writing in order to discover the techniques that work best for you. This chapter will describe a range of tools and approaches, but you should think of these as starting points only and should evolve personally successful variations. Keep in mind that the main—and often the only—reader of such writings will be you, the writer, so feel free to experiment.

A word of caution: Law enforcement officers should check with supervisors about how much of the writing related to an investigation is considered *discoverable;* that is, open to scrutiny by attorneys and usable as evidence. For example, in some jurisdictions the notes an investigating officer makes during an interview or search may be used as evidence, so extreme care should be taken in noting accurately and in summarizing those notes after the fact (see Writing and Memory: Taking Good Notes).

WRITING AND MEMORY:
TAKING GOOD NOTES

When they listen to lectures or discussions, people often regard note taking as a race. For fear of missing something, they try to scribble or type as fast as possible. Not only do they get fatigued quickly, but they miss most of what they tried to hear.

Effective note taking should usually be carried out in at least two stages:

1. *Quick jottings of key words or phrases* during the course of a lecture, discussion, interview, or search.
2. *Summarizing soon after the event,* while memory is fresh and the jottings can spark fuller recall. The goal of this summarizing is to organize the information in some meaningful way, perhaps chronologically or according to greatest significance. This step is especially critical as a follow-up to investigative interviews so that important details are not lost or distorted.

Note: See Chapter 3 for specifics of taking notes during interviews. While good summarizing after an interview is critical, an officer's note taking *during* an interview should be kept to the minimum. For reasons of personal safety, attention should be kept on the person being interviewed, not on your notepad.

3. A third stage, *revision,* may follow if you are reporting the lecture or meeting to another reader. (See Chapter 2 for ways to make all types of revision productive, and Chapters 3 through 9 for ways to revise specific types of writing, from reports to memos to speaking outlines to essay exams.)

You should think of notes also as a basis for further dialogue with a speaker or discussants. It may be useful for you to assume the role of the *investigative journalist,* who not only is interested in accurate recording and useful interpretation, but also intends to follow up with questions about provocative, vague, or seemingly contradictory statements.

To this end, the jottings you make during an event should reflect *both* your interest in capturing the main concepts of an event and your observation of what seems puzzling or inconsistent or unclear. Frequent use of the question mark in notes lets you keep track of concerns that demand follow-up dialogue.

Tools

Technically speaking, while paper and pen/pencil remain the technology of choice for most note takers, especially away from the classroom or office, laptop computers are becoming more frequent in lectures and meetings. As long as they can be used comfortably and without keyboard noise distracting those speaking and listening, computers provide the advantage of facilitating revision of summaries into reports or other final versions for other readers.

Using Visual Aids, Body Language, Prewritten Outlines, Agendas

Jotting occasional key words during an event allows you to attend to the visual dimension of a lecture or discussion, which may be just as meaningful as the heard portion. Body language and facial expressions may give at least as significant an indication of emphasis or doubt as what is said, and a writer may reflect this information in notes, too.

If a lecture or panel includes any material via overheads or computer projectors, ask the speaker about access to those materials *after* the event. Since the visuals usually indicate a speaker's emphases, having outside access can save a writer much note-taking time and will guarantee greater accuracy.

The same goes for preprinted agendas or program outlines. These tools can guide note taking because they indicate some of the speaker's emphases and how the speaker has organized the information. An outline sheet can be used as a kind of *template* for notes; you can elaborate on outline topics in the space available.

The Post-Event Summary

Writing up notes and your other observations soon after an event allows you to use short-term memory. The value of this exercise can't be overestimated, given the swift loss of most of these memories. Note taking during an event aids in retention but is no substitute for a concentrated writing exercise like the post-event summary. (See Chapter 3 for the role of note taking during the verification phase of the interview.)

The most straightforward method of writing up notes merely elaborates on the *chronological flow* of an event. During writing up, a typical note, "nonverbal deception," might become "poor eye contact and shifting feet while answering question." This kind of summary is no organized document, merely a fuller record of your observations. Nevertheless, it can serve as a substantial basis for later study or adaptation into another document.

Alternatively, carefully reading over your notes and thereby reliving the event can give you a sense of a *center* or *focus* for your summary. Trying to find such a center is what the experienced news or sports reporter does after witnessing an event. This center is then reflected in the reporter's *lead paragraph*. The rest of the details of the event are ordered according to how they contribute to this central idea.

Be careful, however, not to jump to an impression of the event that isn't thoroughly substantiated by actual observation. It's basic to *read your notes over thoroughly* before you determine a central idea.

WRITING TO IMPROVE READING: MARGINAL AND FINAL COMMENTS

Writing in, on, and about printed texts is a tried-and-true method for *remembering* and *thinking critically* about anything you read. If you are working on a book or xeroxed copy you own, always be ready with a pen or pencil to jot a comment or a question in the margins: top, bottom, or sides. Using this method of annotation is far superior to the popular—because too easy—method of applying yellow highlighter to any passages you think might be significant. *Writing* a comment or question gets the brain involved in thoughtful reading to a degree that mere highlighting can't approach. Besides, writing lets you express a wide variety of ideas and concerns about the text at hand; highlighting merely highlights. Sometimes readers unsure about which portions of a text are more significant than others highlight whole paragraphs, only to find on rereading that the highlights have not aided their retention and certainly provide no new information.

As with note taking on lectures and discussions, text annotation works best if you regard yourself as engaged in *dialogue* with the author of the book, article, or report. While certainly in most cases a reader can't easily make the comments and questions known to the author, writing those comments and questions forces you to *focus and articulate* your thoughts about an idea you read. These annotations can therefore (1) serve as the basis of ideas that you can develop into a formal piece of writing, and (2) help you and others remember later *why* you found the passage worth marking.

Making Annotation Effective through Critical Questions

Skillful readers read with one or several purposes clearly in mind because a definite purpose strengthens concentration. For example, a software manual reads better if you try out the software immediately and need answers to specific questions about installing and using the product. You read fiction more skillfully if you try to compare features of the author's style to those of another favorite novelist. The clinical psychologist will read a case study more carefully if she has an eye toward understanding a patient with whom she is working. Certainly, the criminal investigator will read an interrogation transcript more skillfully if he or she is looking for signs of deception or omission, and so forth. Even the student fulfilling a textbook reading assignment for a teacher can read with strong purpose, hence effectively, by applying well-known techniques.

If your specific purpose for reading is not already clear, the most common way to give your reading purpose is to *ask specific questions* to guide your reading and annotation. Here are some examples:

1. *How would I summarize this reading for a person who has not read it?* The goal of this question is understanding the *main concepts* in the work—both stated and unstated. As you read a deposition, for example, consider how you'd answer a fellow investigator's questions: "What is the deposition about? What point is the writer making explicitly? What does the writing imply? What is its main evidence? What important information is left out?" Use the margins of the page or a separate sheet (or a computer file) to take notes toward building answers to these questions. For example, observe these annotations by an investigator on a public written statement made by an official suspected of corruption:

No Admission of Guilt

How could anybody say I'm an example of long-term, engrained corruption in our city? I started out operating a pushcart on Market Street selling fruits and flowers. With hard work I parlayed that into several carts and in a couple of years we had cornered the market. Nobody could match our product or prices. From there it was a matter of branching out. Times were tough and a lot of businesses were in trouble and looking to sell. I picked them up for a song. A dry cleaners, a liquor store, and so forth. I got into local politics because I had an interest in what happened there. I never intended to wind up as revenue commissioner.

Review income and loan records to substantiate or refute.

Check to identify his associates; i.e., who is "we"?

Review records for vandalism, accidents, etc.

Determine dates and identify major supporters.

No sense looking for admissions; going to be rags to riches story. Instead, look for details that can provide avenues of inquiry.

2. *In what ways does this writing comment on issues or topics that are important to me?* Even if the piece you are reading *seems* unrelated to a case or project in which you are interested, this question will productively test your imagination. Using this question, the reader watches for statements that bear on topics in which the reader does have some intellectual or emotional investment. Marginal annotations can track these connections. Writing can be especially valuable as a tool to forge connections where there don't appear to be any—and thus where the reader might have trouble feeling motivated to continue reading. For example, note this annotation by a robbery investigator of a general account on plant security:

Could this reference to false alarms help explain why nobody paid much attention to the unlocked safe at Allied?

3. *How can I simplify the language of this piece for someone who doesn't have a clue about the jargon?* Using this question as a guide will push the reader to get into and through technical terminology or insider lingo to a better understanding of the concepts and details. For example, see how this transcript of remarks made by a potential witness was annotated by an investigator who wanted to clarify the document for a new prosecutor unfamiliar with the informant's slang:

As sure as my name is Little Mack, I'm telling you the young thugs with no respect are moving in. Already they control the line and layoff and expect a double dime for the service. You know why? Because the books have been closed so long there is no competition from us. The man is finally pushing for guys to make their bones but it may be too late. Everybody will be taking to their mattresses before it is over.	*services vital to bookmakers* *$200* *organization has no young members.* *organization boss* *qualify for membership* *pending gang war*

Based on the foregoing, the investigator might tell the prosecutor: "Our source recorded the remarks of a Mack Johnson, a known gambler and mob associate who said that young outsiders have moved into the gambling business. Johnson said the mob is taking in new members to combat this, and he is convinced that a gang war will result."

Likewise, academic reading often remains opaque because readers don't make a concerted effort to penetrate the language. Readers often give up on difficult technical material, usually resigning themselves to their own inadequacy or to criticizing the author's style. Marginal writing forces the reader to slow down the pace of reading and strive to understand terms before moving on.

WRITING TO IMPROVE READING: KEEPING THE READING RESPONSE LOG

While marginal annotation allows you to keep up a running critical dialogue with a writer, the *reading response log* provides room for extended (or brief) reflection and analysis. Moreover, by its nature, this ongoing collection of *dated entries* allows you to observe changes in your own perspective and growth in knowledge. Indeed, later entries that summarize and interpret this progress commonly characterize well-kept logs.

Tools

Though the word *log* usually sparks an image of a large, hardbound book, more and more writers keep logs on computer. While the traditional book has the advantage of portability (an advantage somewhat minimized by the laptop), the computer provides the writer ease in (1) *expanding and revising* a strong entry or series of entries into a formal article or report, and (2) *sending* an entry to a fellow reader via electronic mail.

Here are some tips for keeping a reading response log:

1. Always *date* entries, including the year. Even if the current entry adds to an earlier one, date the addition.
2. In referring to specific passages in a book or article, always take the time to *cite the work accurately and fully*, with enough detail so that you or another person could *easily* find it again later. Though log keepers usually want to avoid delays in getting their thoughts about a piece of writing on to the page or screen, making accurate citations eliminates the later frustration of being unable to locate important passages.

Techniques

Think of the log as being as flexible, experimental, or structured as you wish it to be. Because the log is a *personal document*—usually for the writer's eyes only—most log keepers appreciate the freedom it gives them to *explore* ideas and *experiment* with different styles and formats. On the other hand, the log also frees you, paradoxically, to impose on yourself the discipline of specific kinds of critical analysis.

On the side of experimentation, the log can be similar to the *commonplace book,* an age-old tool of scholars and writers for collecting and commenting on interesting texts that one encounters in the course of study or casual reading. The *writer's notebook,* in which professional writers try out imitations of style and collect impressions of events and people, is another experimental option for the log keeper.

On the side of greater analytic routine is the *scientific notebook,* that precise tool of the practicing scientist. In this type of log, the scholar/writer observes and records phenomena according to strict methods of the discipline. Indeed, many scientific notebooks cross the line from the private to the public document because they can serve as legal documents in cases of patent or copyright dispute or in cases of alleged fraud or incompetence.

While we don't suggest that reading response logs should be taken so seriously in most instances, the log can be a good place to impose on yourself a regular analytic regimen. For example, one or more of the three basic analytic questions listed in the previous section can serve you as stimulus for writing about reading. To be more specific: let us say that you are doing research for a report on a complex file for a supervisor or new case officer; you are reading a series of documents on the subject. In the log you can apply question 1:

> *How would I summarize this reading for a person who has not read it?*

to each source read, and this series of summaries might be easily adapted into the research report.

Similarly, you might also *create critical questions* to use as a regular framework for writing about sources. For example, because it is common in scholarly research to question the credibility of an author, standard questions for the log might be:

> *What are the author's qualifications on this topic? What are his/her experience and credentials? What sources does s/he cite to substantiate any claims made? What other evidence is presented?*

Alternatively, a log keeper might want to make a *comparison* between and among the various sources to find a standard regimen for the log entries. For example, let us say that an investigator is attempting to find consistencies and disagreements among statements made in a series of interviews. The officer might begin each entry as follows:

> Informant #1's account agrees and disagrees with Informant #2's version on the following points.

As the sources and, therefore, comparative entries accumulate, the writer builds an overall, integral sense of how all the sources contribute to the officer's understanding of the case. As we emphasized earlier, the questions you impose on your log should follow directly from your carefully thought-out purposes for the reading.

WRITING TO IMPROVE OBSERVATION

Though every discipline relies on a person's ability to read critically, disciplines also demand the ability to observe and interpret other sense data.

The sciences and social sciences obviously demand the researcher's keen attentiveness to sights, sounds, smells, and the like. But the arts and humanities no less require a rich ability to perceive sensually and articulate those perceptions, as this talent is at the center of artistic creation and judgment. Writing can be a tool both to sharpen and diversify your ability to perceive and understand sensual stimuli.

This chapter has thus far described three principal techniques for using writing as a tool of critical thought: note taking, annotation of texts, and keeping a reading response log. The section on note taking mentions that notes should attend not only to *what* is said in lecture or discussion but also to *body language*—and, by extension, other observed phenomena, such as facial expression and tone of voice. Both note taking and log keeping can be easily adapted to the needs of the observer of other phenomena. In this section of the chapter we will suggest a few writing exercises that can make these tools especially effective in these kinds of observations.

Writing Dialogue: An Exercise for Listening

In this exercise, your task is to record as closely as possible *exactly* what is spoken in a conversation and *how* the words are said. You might use a tape recorder and then transcribe the tape, but the important act in the exercise is to write as accurately as possible. In striving for exactness, pay attention to what you hear—rather than, as listeners usually do, trying to penetrate beyond or "inside" the words for what you might either want or expect to hear. Such an exercise is valuable training for any number of disciplines. Some obvious examples would be medicine, law, law enforcement, nursing, social work, and psychology, where accurate listening is a necessity. Another obvious example is writing for theater (or film), in which the realistic reproduction of human speech is essential.

Close Description of Objects: An Exercise for Viewing

As with listening, seeing is often hindered by the human tendency to "correct" what is actually before the eyes toward what one desires or expects to see. To illustrate this tendency, one of us routinely asks students early in a course to "draw a face." Almost without exception, the students look only at the page before them and draw their *idea* of what a face looks like. It is the rare student who in fact looks at another person's face and attempts to draw what he or she sees.

In this exercise, you are to contemplate two or more very similar objects—apples, for example. Then describe in as much detail as necessary one of the objects, so that *another person could identify that object from all the others just on the basis of the description.* Not only does the writing force close regard of the objects, it also challenges you to discover a range of *descriptive devices*—color, shape, statistical measurement, metaphor—to make the fine distinctions.

Describing a Process: An Exercise in Narrative

Here you will attempt to describe a brief process so that a person unfamiliar with the process can perform it. Such exercises, of course, are at the center of the scientific endeavor, since a basic purpose of reporting experiments is to enable other researchers to perform the same experiment. But *process description* is equally central to teaching the methods of every discipline; moreover, business success and effective government depend on the ability to communicate clearly *instructions* for everything from filling out tax forms to taking pain remedies to working a video camera.

Frequently practicing process descriptions in writing will teach you:

1. *Respect for different readers:* How technical can my language be? How can I inform the novice without sounding condescending to the more knowledgeable?
2. *Close attention to even the smallest step in a process:* Testing process descriptions usually reveals that the writer has overlooked small steps that the writer performs unconsciously; for example, one of us repeatedly failed to teach his daughter how to drive a car with a stick shift until he closely compared how he had *told* his daughter how to disengage the clutch versus each small step in how he actually did it.
3. *Diverse ways to "bring alive" a narrative:* Writers learn to give vivid examples; create interesting characters; use graphics, metaphors, humor, and irony.
4. *Respect for feedback and revision—the "process" of writing:* The test stage of a process description almost always teaches writers that they must rewrite before their description can meet a reader's needs.

Here is a sample process description. Observe how the writer has limited the use of technical language, tried for precise accuracy, and used some of the devices noted in item 3:

> The message on the top of the "childproof" aspirin bottle reads: "To open, line up arrows on cap and bottle. Push cap up with thumb." Anyone who's ever struggled with some of these childproof caps knows that "push cap up with thumb" leaves a lot unsaid. More precise might be:
>
> > *Line up arrows on cap and bottle. Insert thumbnail in narrow space between cap and bottle; push.*

Better yet:

> *Line up arrows on cap and bottle. Insert thumbnail in narrow space between cap and bottle. Push with enough force to pop up cap without spilling contents of bottle.*

For those with not enough thumbnail to wedge off the cap—or those who don't want to risk a broken nail—how about this?

> *Line up arrows on cap and bottle. Grasp bottle firmly (for really tough caps, you may have to use two hands). While exerting pressure on the opposite side of the cap with the forefinger, push the ball of the thumb hard, exerting upward pressure against the arrow on the cap.*

To this might be added:

> *If you are taking this pain reliever for headache, you may notice that your headache has worsened by the time you get the cap off.*

WRITING TO EXPERIMENT WITH STYLE AND FORMAT

A great way to learn from writing and have fun while learning is to experiment: to play with language, point of view, the appearance of the document, and the many other variables that go into writing. Up to this point, this chapter has emphasized using a log as a way to practice using certain critical questions as a way to make reading and other observations more meaningful. Here we suggest that an equally important function of writing, whether in a log or in separate exercises, is to do things differently from writing to writing.

While later chapters will present some standard *formats* for documents in law enforcement, keep in mind that the potential for writing in any discipline extends far beyond the standard or customary. Indeed, you can't bring your ideas to some readers without trying out designs, "voices," metaphors, and the like that might strike some as strange or even wholly new. For example, just a few years ago the idea of composing business documents with sound and video would have seemed like science fiction. But now, thanks to the enterprise and imagination of a few, Internet-based business is a booming industry.

Experimenting with New Technology

Writers have always had at their disposal a wide variety of options in style—essays, stories, poetry, song, and drama are a few of the better-known ones. Now, however, with widespread access to sophisticated graphic and presentation software, the average writer has choices that formerly belonged only to a handful of skilled artists and craftspeople. For example, the writer who, regardless of formal training in art, works in the medium of the Internet not only can manipulate the printer's palette of font, sizes, color, and arrangement, but also can adapt the most sophisticated graphic—even cinematic—effects by importing material from other sites on the Net. A writer who creates an electronic folder of stylistic experiments is in fact creating an electronic log that teaches in the doing and that can be used as a sourcebank for other projects.

Necessary Experimentation: Reaching New Audiences

Even without employing computer technology, the writer in any discipline can productively and enjoyably experiment with style and design. An excellent way to get started is to imagine how your language, layout, information, tone of voice, and other features would have to change if a *radical shift in audience and purpose* of the document should occur. For example, here is the cover page of an investigative report (you will see this cover page again in Chapter 5); as you read, think about the writer's purposes in the document and about the characteristics of the typical reader. (You might try annotating the text toward understanding purpose and audience.)

XYX Police Department

Case Number	Burglary—1234–97
Case Officer	Detective John Smith
Inclusive Dates of the Report	xx/xx/98–xx/xx/99
Primary Suspects	William James Joseph Johnson
Victim	Neighborhood Computers, Inc. 555 Maple Street (Merriwether Plaza)

Summary: On xx/xx/98, the above store was burglarized sometime during the early morning hours. Numerous printers, computers, and other hardware valued at over $30,000 were taken. Suspects apprehended three months later while trying to sell some of the proceeds to an undercover officer in a sting operation. Both suspects have been charged with grand larceny and are in custody awaiting trial. Others are suspected of being involved, and investigation is continuing.

Distribution:
1—Burglary Squad Supervisor
1—District Attorney
1—Case File

Now imagine a dramatic shift in reader and purpose: for example, the reader is a small child and the purpose is to keep the child from being terrified about the crime, but also to impart a lesson in safety; or the reader is a person interested in a secure financial investment and the purpose is to convince the person to invest in Neighborhood Computers, Inc.

In experimenting with changes in style and format, consider some of these categories:

Level of technical language—what words need to be defined? for which terms should substitutions be made? Should some ideas be eliminated altogether or spoken about only in metaphor?

Tone—play with mood: threatening, carefree, cryptic, earnest, fanciful, professional/technical, reassuring, etc.

Syntax—short, simple sentences? long, complex ones? questions?

Person and voice—personal and active ("I'd do X and so should you")? impersonal and passive ("research was conducted which might suggest . . .")? Consider how person and voice can affect readers' respect for you and their interest in what you have to say.

Use of story and character—"Imagine walking into a big room that smelled like . . ."; "Von Delbach stumbled on this technique when she was working on . . ." Story and character usually increase a reader's attentiveness.

Conventional paragraphing versus frequent indenting, use of "bullets," bold-face type, shifts in font, size, etc.—Imagine, for example, how different this page would look if all the information were printed in two or three paragraphs, with no spacing, no headings, and no use of italics.

Here are two stylistic experiments based on the sample investigative report cover page:

1. Audience: Small child. Purpose: impart safety lesson without terrifying child.

Oh, now don't be scared. You're all right now, aren't you? The police caught the men that robbed the store. We go shopping all the time, don't we, and we don't see any robbers, right? So don't worry. Now, what did we tell you about holding onto Mommy or Daddy's hand when we're in the store? That's right. We just want to keep you safe, because we love you so much.

2. Audience: potential investors. Purpose: encourage investment in small, growing business.

We are pleased to report that _____ police have recovered more than half of the $30,000 in equipment stolen during the robbery on xxxx xx that was reported in the local press. The police investigation is continuing, so there is opportunity for further recovery of property. Neighborhood Computers is fully insured for any loss.

Two suspects were apprehended and have been charged with grand larceny. As reported in the press, the men in custody are also suspected in recent robberies at other technology stores in the metro area. Detective John Smith, chief investigator of the case, praised store management for its cooperation in the investigation. He noted that Neighborhood Computers' store security policy is primarily responsible for the building's not having been robbed at any previous time. That Neighborhood is located in Merriwether Plaza, which police statistics show is the safest mall in the metro area, will further ensure that any future burglary is highly unlikely.

GOING PUBLIC: FROM WRITING FOR YOURSELF TO WRITING FOR OTHERS

Chapter 2 will describe techniques for transforming the writings (annotations, "working papers") you do for yourself into documents meant for others. "Going public" with writing has often stopped would-be writers, who fear the scorn of readers. Keep in mind, however, that if you regularly practice using the tools and techniques described to this point, confidence will increase as your skills grow and diversify. With practice tools such as the log and a ready bank of questions for critical reading, seeing, and listening, there is no reason why a writer should ever be stymied by a blank page or screen.

2

THE WRITING PROCESS: PREDRAFTING, DRAFTING, REVISING, EDITING

GENERAL PRINCIPLES, BUT NO SINGLE FORMULA

In its details, the writing process varies for every person. Even for the same writer, the process he or she uses will vary somewhat from task to task. When writing specialists talk about the writing *process,* they mean several things: for example, the steps and the attitudes through which a writer proceeds toward completing a writing task, whether it be a memo to a co-worker, a research report to a professor, or a letter to a relative. Understanding how writers accomplish such tasks helps writing researchers improve the teaching of writing and the development of teaching materials.

To the composition scholar, the writing process also means a general progression that is exemplified over and over in the work of experienced writers, regardless of field. Though scholars have given differing names to these stages of the writing process, they are most commonly referred to as "prewriting" (in quotes because we will be substituting what we feel is a more appropriate term), *drafting, revision,* and *editing.* Each stage will be dealt with in this chapter. Chapters 3 through 6 (four stages of "Writing the Investigative Report"), Chapter 7, Chapter 8, and Chapter 9 follow the principles of drafting and revision described here.

"PREWRITING" AND DATA COLLECTION

Like anything else worth doing well, writing requires *planning* and investment of a good bit of time *before* you sit down to make what will become the finished work. In the case of writing, this means that you should plan to do some, often considerable, writing before you actually draft the paper, report, or proposal. Chapter 1 describes such writing tools as the log, which can be very valuable in this planning phase. Depending on how it is structured by the writer, the log can help you

1. Plan the *content* and *organization* of any written work.
2. Sort through and *evaluate* readings and other data you plan to use.

As described in Chapter 1, the log can also be a place where you *experiment with style and format* for the project being launched.

The work you do before drafting is often referred to—somewhat misleadingly—as "prewriting." A more appropriate term would be *predrafting*, since, as we've seen, much writing is often involved in the planning stage of any project. Indeed, if you use this planning—predrafting—stage for log keeping, as described in Chapter 1, some of what you write experimentally in the log may indeed find its way into the actual draft.

The keys to successful predrafting are *patience* and an *open mind*. The more complex a writing task, especially the less knowledgeable you are about the subject and format of the project, the higher percentage of the total project time you will need for predraft work. Don't be surprised to find that 80 percent or more of the total time you spend on a project will be devoted to the predrafting stage of, say, a piece of original research. Writing during this stage, perhaps kept in a log (as noted) might include:

notes on lectures, interviews, conferences, or experimental procedures

annotations in books and on articles

summaries of notes and annotations

experiments in style and format

Patience is vital in allowing this rich process of reading, talk, writing, and analysis to shape the ideas that will be central to the draft. (See Chapter 1 for more detail on these techniques.)

An open mind is similarly essential. As Chapter 1 suggests, the purposes of the predrafting stage are *learning* and *critical thinking*. Writers who, because of haste or impatience, move too quickly to the drafting stage, jump to opinions about data and then feel constrained *by the draft* from collecting more evidence and from doing fresh thinking. *Budgeting* sufficient predraft time and *using critical thinking tools* such as the log almost always ensure a better finished product, but writers need to be patient and willing to learn in order to be comfortable with these aids to writing.

Doing a Dummy Draft

One experimental technique that can be part of the predrafting stage is what we call a *dummy draft:* not really a draft that incorporates all of your note taking, summarizing, and other preparation, but an attempt during your predrafting to *set down in an organized way what you have learned about the subject so far.* This dummy draft can be part of the log. Its purposes are:

1. To *bring together the many details* you have accumulated during predraft research in order to let you see what you've accomplished to this point.
2. To *reveal the gaps, flaws, and inconsistencies* in the research and thus indicate what still needs to be done.

Like the dummies constructed by engineers in the design process, this draft allows you to *test out* ideas and wording. The dummy draft is informal, its only reader the writer (and perhaps a close colleague who has agreed to give feedback on the preliminary work). This piece can help satisfy the writer's anxiety about the sufficiency and significance of the research to this point. It can also be *concise,* just long enough to achieve the two purposes just noted.

A word of caution: As noted at the beginning of Chapter 1, law enforcement personnel need to know policies in their jurisdictions about how much written material is *discoverable*—that is, open to scrutiny by attorneys and usable as evidence. Some jurisdictions distinguish between discoverable notes and reports and so-called *working papers,* such as drafts of reports, which are not open to scrutiny. Though the dummy draft is not a full-scale draft of the report, it is part of the drafting process and so should be classed as a working paper.

To show something of the flavor of such a work, on page 19 is an excerpt of a dummy draft of an article by one of the coauthors about the Sherlock Holmes stories of British author Arthur Conan Doyle. Note how he uses the dummy draft to write questions and observations about the writing.

Predrafting and Specific Writing Tasks

The following chapters will briefly describe appropriate predraft writing for such specific writing tasks in law enforcement as the interview report, the search report, and the internal memo. In general, though, our advice is

Sample Dummy Draft

Doyle was not the first writer of crime fiction to cast the police as foils for his heroic detective. More than a half-century earlier, Edgar Allan Poe had contrasted the systematic and thorough, yet unimaginative and ineffective, methods of the Paris police to the incisive brilliance of Monsieur Dupin in "The Purloined Letter." (Check out any explicit mentions of Poe's influence by Doyle.) Even three centuries earlier, the versatile innovator Robert Greene, in his London "conny-catching" (con game) stories, had created eloquent rascals who scoffed at the plodding efforts of the constabulary to detect their scams. (How far could I take this back? The devil figures in the morality plays? The wily servants in Roman drama?)

The police in these stories are always portrayed as severely limited, usually in intelligence and sometimes in character. How the police are portrayed depends on the writer's purpose. For example, Greene's "watch" or "sheriff's men" (check exact terminology) have no personality or power at all; his flashy villains carry out their cons, burglaries, and "purse nipping" at will. If they eventually fall to the officials, it's only because jealous rivals have informed on them. (verify) This invisibility of the police suits Greene's purpose, which is to show that evil begets evil: crime pays for a while, but inevitably the criminal is destroyed by his own perversion or by the anti-society it creates. His focus is on the soul of the criminal; the police are irrelevant to his purpose. (It would be interesting to see if in Greene's time there were any concept of a "professional" police who were supposed to catch criminals through detection methods.)

Conversely, Doyle's police play a key role in his characterization of Holmes as detective par excellence. Doyle's police fall into two general types: those stupidly arrogant enough to believe that they can outwit Holmes and those smart and humble enough to call in Holmes to save the day. Either is a perfect foil. Both types share positive traits that Holmes readily acknowledges: they are honest and respectable, they work doggedly, and they do the best they can. Even the arrogant ones, epitomized by Inspector Lestrade of Scotland Yard, are model citizens. By contrast, Holmes, though impeccably patriotic, cuts a crooked figure. Manic-depressive, abusive of cocaine and tobacco, given to long silences, disappearances, amazing disguises, and explosive bursts of energy, Holmes defies commonplace respectability. The contrast shows Doyle's purpose: only Holmes can contend with the great criminal minds because only Holmes can see, as they do, how to manipulate reality and have the courage to do it. By implication, the stolid decency of Doyle's police is of a piece with their inability to see beyond their routine techniques and hasty, simplistic judgments. (Where, if anywhere, does Doyle, through Watson, come out and say this? Holmes frequently criticizes the severe limitations of police methods and imagination, but does he ever say anything like "They are too decent to be good investigators"?)

the same, regardless of the task: sufficient time spent on such working papers as the log and dummy draft will pay off in better and more easily written documents.

DRAFTING: A CHANGE IN ATTITUDE

Moving from the predraft to the drafting stage involves a change in attitude. The most useful attitude to take toward the varied writing in the predraft stage is *experimental;* the main audience is you, the writer, and the primary purpose is *learning.* When the writer drafts in earnest, the most useful attitude is still experimental, since the writer must feel that whatever is being written is only being tested out, and can be revised. However, the sense of audience has changed: now the writer is directly attempting to reach other people with definable characteristics as readers and definable needs for information.

The basic *purpose* has changed, too, from the writer's own need to make sense of reading and other phenomena to *meeting the needs of those defined readers.*

These changes mean that the drafting writer writes with a double consciousness: you write *both* with an open mind, that is with an eye toward possible revision, and with a sense of *limits* imposed by the purposes and the possible readers. Thus, the drafting writer has a very focused intent: "I'm writing this the way I want it to look to my intended reader"; yet you also feel the ease of knowing that this is not a final draft; hence, you can ask for *feedback*—critical commentary—on this draft that can lead to productive revision.

How Rough Is the Rough Draft?

Unlike the dummy draft that you write during predrafting, a so-called *rough draft*—another kind of working paper—should be intended to *stand as a final product,* unless feedback dictates that revisions are needed. In either academic or workplace situations, when teachers or supervisors ask to "see a draft," they almost always mean a seriously written effort that incorporates the writer's best thinking and most thoughtfully analyzed evidence. A rough draft is like a rough diamond: not polished, but still a diamond.

That's why the so-called "rough" draft comes *after* the intensive work of predraft writing and research. Inexperienced writers often make the mistake of putting off the hard work until after the rough draft, which itself appears shoddy and which comes too late in the process to allow for the careful study and critical thinking that should have occurred in predrafting.

PLANNING THE DRAFT: THE THREE KEYS

During the predraft stage, writers should of course be guided in their study and planning by their thoughtful understanding of the purposes of the project. As illustrated in Chapter 1, the writer who has a clear sense of purpose will learn the most and think most productively about any subject.

Also useful in the predraft stage, but absolutely essential toward writing a solid draft, will be clear understanding of two other factors: *format* and *audience*. With *purpose*, these are the three keys of good draft writing.

Formatting the Draft

The format comprises both the *organizational structure* and the *appearance traits* of the draft. Chapter 5 will describe efficient formats for investigative reports; Chapter 7 will show formats for different types of memo; Chapters 8 and 9 will do similar work for oral presentations and essay exams, respectively. Nevertheless, be aware that it is hard to generalize about format, because the "right format" for any given document will probably include two features:

1. Highly specific requirements in the particular situation.
2. Leeway for the writer's creativity.

Learning the Highly Specific Format Requirements

If you are responding to an assignment from a particular reader—say, a professor, supervisor, or case officer—*assume* that there are specific format requirements and *seek* to learn them. Don't be misled by the assigner's silence about format; he or she may be assuming that you already know all the tiny rules. Then again, the assigner may have never given thought to what s/he expects—such readers often say things like, "I'll know what I want when I see it."

There are two basic ways to *seek* the specific formatting:

1. Ask (see the checklist of format characteristics).
2. Study and emulate earlier documents for the same purpose and the same reader. For example, grant agencies usually make available on request successful previous applications. Businesses keep files of their documents, which can be used by new employees as models of style and format; professors often do as well.

Caution: Get the opinion of the supervisor who assigned the task before relying on any earlier file as a model. Files usually contain bad models as well as good. Show samples to the supervisor and let him/her choose the best to follow.

When you *ask* about format, the following checklist of typical format characteristics may be helpful:

> *Correct spelling, punctuation, and Standard Edited American English (SEAE) syntax:* Assume these are required unless otherwise informed.
>
> *Order of information:* Should the draft be organized in sections? If so, in what order?
>
> *Use of headings and subheads:* Should sections of the draft have headings or titles? What headings should be used?
>
> *Margins, spacing between lines, and indentations:* Are there specific rules for these?
>
> *Boldface type, italics, underlining:* What sorts of items should receive these kinds of emphasis?
>
> *Number of words or pages, minimum and maximum*
>
> *Type style (font) and size*
>
> *Illustrations, photos, charts, and graphs:* Are these required? frowned on? Are there size and style restrictions?
>
> *"Special effects"—video/audio:* Can video or audiotapes accompany a written draft? How fancy should a Web page be? If these effects are used as addenda, how should they be referred to in the written text?
>
> *Cover pages or cover letters:* Should these be used? If so, what information should they contain, in what order, and with what appearance?
>
> *Footnotes, endnotes, and citations:* Is a specific documentation style favored? How do the writers of model samples cite sources?
>
> *Appendices and other addenda:* Are these allowed or encouraged? If so, are there page limits? How should data in the appendices be cited in the main text?
>
> *Quality of paper or other materials*

Formatting on the Basis of Models

If the writer can follow earlier models in formatting the new document, the checklist of format characteristics may also be helpful in analyzing the models. Do not hesitate to ask the intended reader questions about formatting, audience, or purpose that may arise in reading the model documents. For example, it's common for model documents to provoke questions about *formality* or *informality of tone*, about the *technical level of the language*, and about the *assumed level of the reader's knowledge*.

Two General Rules of Formatting

In the absence of specific information from a reader or of sample documents to use as guides, two all-purpose formatting rules of thumb can be used:

1. *Be simple and consistent.* Most readers respond well to clear layouts and consistent use of spacing, indentation, and other features; frequent shifts in font and type size tend to distract and confuse.
2. *Format so that readers can grasp your main ideas as quickly as possible.* Judicious use of headings, spacing, indentation, and emphasis (boldface, italics) guides the reader's eye as the writer wishes. Long paragraphs, tiny type, and little white space tend to confuse and (dare we say it?) bore.

Audience: Drafting for Your Reader

Unless you write only for yourself, you write for more than one audience: yourself and at least one other person. And even the audience of the self changes depending on mood, fatigue, and your latest experience. So drafting for your audience is no easy task, and it becomes more difficult the more readers who are involved.

Of course, the more practice you get writing for certain readers, the more you can assume about them and the easier the task becomes. But whenever the writer, no matter how experienced, tries to reach a new reader, some planning is called for.

As with learning about format, learning about other traits of audiences may require *asking readers* and *using model documents.*

Talking with Readers (and Writers)

Don't hesitate to request further information about new assignments from teachers or supervisors. Some common concerns are those noted earlier:

Tone: How businesslike? Reserved? Official? Impersonal? Friendly? Solicitous? Glib?

Level of technical language: Which terms should be defined? Should any terms be avoided?

Assumed knowledge of topic or issue: Should I assume that the reader has also observed the scene? Does the reader already have a particular viewpoint on this case?

By all means, be sure to ask readers if they would be willing to give feedback on a draft of the document. Receiving such an invitation may be a writer's best means of ensuring a high-quality final draft. (See the Revision section of this chapter.)

Also don't hesitate to talk with writers more experienced in performing the kind of task you've been given. Even the simple question "What should I look out for?" will usually provoke an informative response.

Learning from Previous Models
Besides using the checklist of format characteristics on p. 22, apply the questions about tone, level of technical language, and assumed knowledge to previous examples.

An Audience Exercise for the Log
Before drafting, write an "Audience Analysis" as part of your log. Use this writing to think about characteristics of the various groups who might read your work. Make some preliminary decisions about tone, technical language, and assumed knowledge.

Knowing the Purpose(s) of Your Draft

The third key to good drafting is being sure of purpose. As with format and audience, *asking readers and experienced writers* and *learning from previous models* can provide insight about purpose. (See the previous sections on Format and Audience for details.)

Be aware that the purposes of any document are multiple. Even a two-line scribbled memo to a co-worker: "Let's talk at lunch about the budget proposal," can have many intents, most of which may be indirect: (1) "Let's talk at lunch about the budget proposal"; (2) "I'm including you and ex-cluding _____ from our talk"; (3) "I'd like to have lunch with you"; (4) "I hope you'll have lunch with me"; (5) "This is handwritten and off the cuff; it's no big deal if you refuse"; and so on.

When talking to a reader or a more experienced writer, show awareness of these multiple purposes. Instead of asking "What is the purpose of this document?" ask "What is the most important purpose? If it does nothing else, what must it accomplish? What are some other purposes it should achieve?"

A Purposes Analysis Exercise for the Log
Before drafting the project, use this log piece to help you list and prioritize the purposes of your project. It may help you discover purposes of which you had not been aware; it may help you decide which information should come first; that is, in the position of greatest importance in most academic and business documents.

EFFECTIVE REVISION

While any experienced writer would agree that "writing is rewriting," making that rewriting, or revision, effective calls for sound strategies, not just good intentions.

First, let's clarify what we mean by revision. Basically, writing specialists define *revision* as:

1. *Changes* that a writer makes to a draft *as* it is being written; as writers compose, they return to earlier portions of a draft and change the text in response to fresh insights.
2. A more *systematic process* by which a writer submits a draft to a reader or readers for feedback, and then makes changes based on reader commentary.

Revision is *not* to be confused with the copyediting, proofreading, or re-copying that writers do before submitting a final draft to a professor, boss, client, or other reader. Such changes don't involve the thoughtful, often imaginative *re-seeing* that characterizes revision.

Why Do Writers Revise?

As part of the normal process of communication, more than one draft of a spoken or written statement is often needed in order for a message to be clearly understood. Inexperienced writers often wonder why writing over which they have exercised great care still fails to communicate. Experienced writers have learned that the need for revision is usually a sign of neither the writer's nor the reader's incompetence, but simply a normal task.

Moreover, experienced writers have also learned to *use* revision as a way to intensify and expand their own thinking. Trying out different phrasings, different organizational patterns, and different ideas changes one's perspective on any task or topic. Using revision in this creative way often leads to a far better result.

Techniques for Effective Revision

Here are some important ways to approach the revision process.

Using Wait Time

Always try to budget some *wait time* into the writing process. The more time away from a draft you are working on or one that you have completed, the better able you will be to see the work as another reader might see it. In the midst of composition, your mind is likely so full of the ideas that you want to convey that it is difficult for you to detect gaps in reasoning or unclear phrasing.

Even overnight may be enough time for you to clear your mind sufficiently to allow for what is called *distancing* from a draft: that ability to see as others might see the writing. Wait time is especially effective in letting you pick up *unclear transitions* from one idea to another and *vague wording*. Of course, if you can set aside a draft for several days before rereading, all the better.

Most writing situations, especially in the workplace, don't allow much luxury of time; however, rarely in law enforcement—or any other field—is a document demanded the same day that it's assigned. Whenever you have any prior notice, or lead time, you would be well advised to draft sufficiently before the deadline to make use of wait time.

Looking through the Reader's Eyes

Using wait time to gain perspective on a draft is necessary for strong revision, but it won't help much if the writer isn't able to see the draft from something close to the perspective that a primary reader of the text would have. The next technique we will describe is how to get good, direct *feedback* from readers themselves; however, since all writers have to be critics of their own work, it's vital that you yourself be able to "be"—or at least imagine yourself to be—those readers.

Some ways to increase your empathy with readers include:

1. *Study documents of the same kind you are composing.* Good writers in any genre are almost always avid, studious readers of the same type of work, whether novels or letters to the editor or crime reports or play reviews. As described in the Formatting section, finding and studying files of similar documents definitely helps in the writing process. If finding such files is difficult, ask the person who assigned you the writing task if samples are available.

2. *Do the Audience Analysis exercise on page 24.* In the log or elsewhere, do one or more writings to help put you in the reader's frame of mind. Consider such questions as "What does my reader wish to get out of this reading?" "What will my reader expect to see first?" "What would be likely to grab my reader's interest?" "What should I avoid in order to keep my reader from getting upset or bored?"

3. *Keep handy the checklists of format, audience, and purpose criteria.* The more you work with lists of criteria, such as those given in the Drafting section, the more you will instinctively apply these criteria. As long as you are still relatively new at working with a particular kind of assignment, it's important to keep at hand your carefully annotated assignment, log writings about audience expectations, and other notes.

Getting Good Feedback from Readers

Especially when they face new types of assignments or write for new readers, experienced writers have learned how indispensable to good rewriting feedback from knowledgeable sources is. Carefully using wait time and keeping handy any written criteria always help produce better prose, but there is no substitute for specific advice from

1. Readers for whom the document is intended
2. Other writers experienced in that genre or with that type of reader
3. Another writer whose opinions you trust to be careful and honest

You should never presume that **Reader 1,** also known as the *primary reader* or *audience,* is inaccessible or unwilling to comment on a draft. For example, although college faculty in most courses do not mandate or even formally invite students to request feedback (though such invitations have become more common), almost all faculty we know are open to such requests from students and appreciate their initiative and seriousness in making the request. Conversely, while magazine editors may not offer to comment on drafts per se, the process by which articles are considered for publication usually means feedback to the writer and the need for revision. Most material submitted to academic journals, for example, is neither accepted nor rejected outright; editors routinely send writers fairly substantial commentary written by themselves and by members of the editorial board, and these responses enable revision by the writer.

Another kind of primary reader is a member of a larger public who will be receiving a document and who will be expected to act in some way on its message. Everything from course syllabi to news releases to advertising is meant for such readers.

Avoid characterizing your audience as "general readers" or the like. Instead, think of each hypothetical reader as an individual with identifiable needs and characteristics. If at all possible, get feedback on a draft from one or more members of that readership. As with the first type of primary reader, give the reader specific questions that will focus commentary and show the reader that you want concrete suggestions—not bland praise.

Reader 2, the *writer with experience in that genre or with that audience,* can be invaluable in pointing out handy tricks and hidden pitfalls. In asking for feedback from such a source, be sure to specify exactly why you are seeking this person's advice (e.g., "I read your report of the Smith investigation, and I know you've had a lot of experience writing these. Will you read my draft and point out some things I should be sure to do and some things I should be sure to avoid?"). Because of this writer's experience, you can rely on his/her being aware of most of the concerns you may have about format, audience, and purpose; this writer will be aware of other concerns that may not have crossed your mind. Be ready to listen, answer questions this writer will pose to *you* about format and other matters, and take careful notes.

Reader 3, the *trusted adviser,* will be valuable to you mainly for this person's candor, the ability to write well him- or herself, and knowledge of how to communicate clearly with you. It usually takes much practice in asking for feedback, perhaps years, for a writer to identify a person with these qualities. Don't confuse a trusted writing adviser with a close relative or good friend. A relative or friend *might* turn out to be a good writing adviser, but it will take time for you to find that out. Never turn to anyone for advice just because they are convenient to you or might feel obligated to give you some kind of response.

Regardless of the reader or the situation, here are some all-purpose rules for getting good feedback:

1. *Always ask specific questions.* Before showing your work to someone, spend some minutes writing several questions that will help your respondent focus commentary. Suggestions: (a) mark passages that you found especially difficult to write, then write out why you are concerned about the wording; (b) ask the all-purpose questions:

"Where do I need to explain in more detail?"

"Where should I cut?"

2. *Make it clear to your respondent that you want constructive suggestions, not a pat on the back.* The least useful question to ask a respondent is "What do you think?" Unless the person knows you well and knows that you really want honest, critical feedback, the respondent's tendency will be to make a generalized "feel good" comment that won't help you revise at all. If you really do need some encouragement—and all writers do—ask the reader what, if anything, is strong in the draft and why. Then move on and ask your specific questions to spark definite suggestions for change.

3. *Listen carefully, take notes, and exercise judgment.* Your attitude to criticism should always be open, but it's also important to maintain integrity: respect for your own judgment. Listen respectfully to every reader's commentary, take notes on and beside your draft, but don't rush to change the draft until you have carefully weighed each comment. Even readers with similar backgrounds will rarely concur on all, even most, suggestions, so a writer should rely on wait time and on his/her heightened awareness of format, audience, and purpose in order to decide how much critical commentary to follow.

4. *If needed, get more than one opinion—but don't shop around.* If you read the Acknowledgments in this book, you will notice that reviews of drafts of this guide have been sought from several people, all of them highly regarded in composition studies or in law enforcement. While it isn't usually practical for writers in most school or workplace situations to ask for a wide range of opinions, strive to avoid giving too much weight to one person's response—especially if that person is other than the professor who will be grading the school assignment or the supervisor who will be judging the proposal, and so forth. Reinforcing comments from several readers will give you confidence in the changes you plan to make.

On the other hand, be careful to avoid mere shopping around for feedback. If the first person to whom you turn is an acknowledged authority on that type of writing, there may be no good reason to consult others. Moreover, consulting others may be interpreted as disrespect for the expert's advice and may jeopardize one's future relations—and not only in terms of writing advice.

Remember that the questions you spend time writing about your draft, format, audiences, and purposes will be the best questions. Nevertheless, here are some all-purpose questions that work in most writing situations:

1. What comes across to you as my main point here? What do you feel I'm mainly trying to accomplish?
2. What do you think I should write more about? What needs further explanation?
3. What could be cut? Why?
4. How would you characterize the person I seem to be writing for? Could you suggest any shift in my sense of the reader?
5. What words or phrases seem to you unclear or misleading? Where are you confused?
6. What questions do you have for me about the draft?

Note that all of these questions must be answered with information, not a simple judgment "yes" or "no," "okay" or "could be better." Your questions should show the reader that you expect to revise and need substantive help in identifying where and how to change.

EDITING THE REVISED DRAFT

Writers differ in their ways of attending to matters of formatting, grammar (syntax), punctuation, and spelling in the drafting and revision stages. Some writers need to keep everything neat and correct as they compose; misspellings, typos, and the like distract their thinking. Others write first, then tidy up the details just before they submit the document to the primary reader. Some do some of both.

Whatever the variations, most experienced writers give their revised draft an *editing review* before submitting it to the reader. This pattern is exemplified in standard book publishing process: the revised typescript is carefully *copyedited* only after all changes have been made in the ideas (content) and organization of the work. Then, once the copyedited text is set into *page proofs* (the actual typeset sheets in the font and size in which they will appear to the public), it is *proofread* at least once more to ensure that all errors have been caught (some are never caught, of course).

This final section of the chapter will deal with a few common errors in grammar, punctuation, and spelling within what linguists call *Standard Edited American English*. SEAE is the specialized dialect of written English in American schools, government, and business. As with all dialects, this one is continually changing, so don't be surprised to see occasional exceptions to these "rules" in some documents.

Three Common Errors in SEAE Grammar (Syntax) and Usage

This chapter in no way substitutes for a comprehensive handbook of English grammar and usage, such as *The Allyn and Bacon Handbook*. This section will only alert you to a few of the most common errors that irk teachers, clients, supervisors, and other readers.

Subject/Verb Disagreement

Rule: If the subject of a sentence is singular, the verb must be singular; if the subject is plural, the verb must be plural.

> **Wrong:** The *Redwings is* the team that represents our town in the tournament.

> **Right:** The *Redwings are* the team that represents our town in the tournament.

> **Wrong:** The full *collection* of books, monographs, and letters *reside* in the university library.

> **Right:** The full *collection* of books, articles, and letters *resides* in the university library.

Vague Pronoun Reference

Rule: In order to avoid confusing the reader, avoid using a pronoun to substitute for a group of nouns not in a simple series. Readers will tend not to know to which noun(s) the pronoun is referring.

> **Wrong:** The Edict of 1245 superseded the Decree of 1218, *which* meant that residents of the kingdom had to register for taxation according to the value of land and livestock.

Right: The Edict of 1245 superseded the Decree of 1218; the *Edict* required that residents of the kingdom had to register for taxation according to the value of land and livestock.

Sentence Fragments

Rule: The subject of a sentence may not be a demonstrative pronoun (*which, that, who*), unless the sentence is phrased as a question.

Wrong: The candidate declared herself a native of Colorado, California, and Illinois. *Which* are the three states in which she lived before going to college.

Right: The candidate declared herself a native of Colorado, California, and Illinois, the three states in which she lived before going to college.

Two Common Errors in Punctuation

Comma Splice or Fault

Rule: A comma should not be used as connecting punctuation between two complete sentences.

Wrong: The company tried TQM during the early 1980s, it downsized drastically in the early 1990s and began outsourcing its training and accounting.

Right: The company tried TQM during the early 1980s; it downsized drastically in the early 1990s and began outsourcing its training and accounting.

Right: The company tried TQM during the early 1980s, *then* it downsized drastically in the early 1990s and began outsourcing its training and accounting.

Lack of the Second Comma in a Nonrestrictive Noun Phrase or Clause

Rule: When a comma is used at the beginning of a noun phrase or clause, a second comma must be used to close the phrase or clause (except when the phrase or clause ends the sentence).

Wrong: *Their Eyes Were Watching God*, a book by Zora Neale Hurston was widely criticized and ignored when first published but has become popular in recent years.

Right: *Their Eyes Were Watching God*, a book by Zora Neale Hurston, was widely criticized and ignored when first published but has become popular in recent years.

Two Common Spelling Errors

Most spelling errors tend to fall into a few categories. Here we focus on two of the most common types of errors.

Misuse of the Spell-Checker with Homophones

While the spell-check function of most word processing programs will catch many spelling errors during the composition process, many writers either forget to use the spell-checker or rely on it to do things it cannot. For example, a spell-checker will not be able to pick up wrong usage of a homophone (a word that sounds like another but means something different). The most common so-called "spelling" errors are actually uses of the wrong homophone (*it's* instead of *its*, *their* instead of *they're* or *there*). So you will need to proofread your documents carefully for uses of these words, even if you use a spell-checker for other purposes.

Hint: If you know that homophones give you problems, and if your word processing program allows you to do this, try *removing from your spell-checker's dictionary* the homophones that plague your writing. In this way, every use of the problematic words will be marked by your program so that you may review the spellings.

Some commonly misused homophones:

it's/its	*their/they're/there*	*flier/flyer*
affect/effect	*discrete/discreet*	*site/cite/sight*
where/wear	*here/hear*	

and a near-homophone: *lose/loose*

The Problem with Spelling by Sound: The Schwa

The most common *kind* of English sound that is misspelled is the *vowel in an unaccented syllable.*

Try saying the following three words: *pennant, independent, lesson.* Notice that the final syllable of each word has the *same* vowel sound, but that a different vowel is used in each case: *pennant, independent, lesson.* This vowel sound is called the *schwa* by linguists, represented in the phonetic alphabet by the symbol ə, a character not present in English spelling. Chances are that if you consider yourself a poor speller, a significant proportion of your spelling errors occur with schwas, because the correct vowel is not determined by sound.

Fortunately, spell-checkers are great at picking up most schwa errors (except in homophones such as *effect/affect*), so use the spell-checker. However, if you are not word processing a document and if you don't trust your spelling, pay particular attention to those unaccented vowels and consult the dictionary.

A Final Note on Proofreading

Proofreading is no fun for most writers. To do it well, you must disregard *what* you have written about and pay *precise* attention to the tiniest details of appearance. Many readers place extremely high importance on correct syntax, punctuation, and spelling; even one error is sometimes enough to ruin the strong impression created by your careful predraft efforts, drafting, and revision. Consequently, careful proofreading is an absolute must for any document with which one wants to win a reader's interest and good will, so time in the writing process must be reserved for this final, equally important stage.

Proofreading for Spelling Errors

If you are not writing on a word processor with a spell-check program, the best way to check for spelling errors is the tedious, but effective technique of *reading your text backward, one word at a time*. In this way, you can concentrate on each word because you will not get caught up in the flow of ideas. But remember to stay alert for misused homophones; remember, too, that this technique only works for spelling—it can't help you catch any other kind of error. (See the Editing section for other tips on checking for spelling errors.)

Proofreading for Errors in Punctuation and Syntax

Refer to the Editing section and keep it handy when proofing your documents. The common errors noted there are often missed by even the most sophisticated grammar-checking programs on computer word processors. Nevertheless, we strongly recommend the most recent versions of such grammar-checking programs, because they are adept at catching such common errors as repetition of the same word (*in in*), too much spacing between words, wrong punctuation, and the frequent failure of writers to remember that a change in one part of a sentence means that they must change other parts of the sentence in order for syntax to be consistent. For example, it's common for a writer to change a plural noun to a singular noun and forget to change the verb correspondingly.

But *use grammar checkers cautiously*. Not only do they miss much more than they pick up, but they also alert the writer to many possible errors that are in fact correct constructions. So tend to trust your own judgment more than the computer's. The best editors are human; in any workplace where correct writing is important, good editors are worth their weight in gold.

3

WRITING INVESTIGATIVE REPORTS: CONDUCTING RESEARCH

PROLOGUE: THE NEED FOR A NEW PERCEPTION

Paperwork! Merely hearing the term causes many a brave police officer to shudder, seemingly with good cause. Horror stories of cases lost and careers ruined because of poor writing permeate the profession. Local, state, and federal agencies all share this attitude; neither an organization's size nor its affiliation seems to matter.

For many investigators, report writing epitomizes this problem, and no wonder. The sheer size of many reports would overwhelm anybody. Just the thought of reading them, much less writing them, boggles the mind. Combined with the knowledge that even minor errors can destroy an entire case, it becomes easy to understand the nearly universal dislike of report writing. Law enforcement must solve this problem if it is going to improve its effectiveness.

The solution must begin with a change in perception. Officers cannot look at the investigative report as a single entity. Instead, they must view it as a compilation of many pieces derived from individual investigative actions such as surveillances, searches, and especially interviews. This viewpoint makes the term *report writing* a misnomer. Officers do not write reports; they assemble them, and that is a far less daunting chore. Once you realize this, it will allow you to concentrate on gathering the facts and recording them during each step of the investigation. The following material is meant to help you do this.

WRITING: REAL POLICE WORK

The keen competition for jobs in law enforcement has allowed police departments to test and eliminate candidates who cannot convey their thoughts in writing. Despite this screening, the lament, "Cops can't write," continues to be heard throughout many departments both large and small. At least in some cases, events confirm this opinion. Incidents, some comic and some tragic, seemingly caused by inept writing by police officers, appear in the news all too frequently.

These criticisms can and have affected subsequent investigations, sometimes causing officers to tailor their findings to facilitate their writing. A story from the 1920s illustrates this: While walking his beat in New Orleans, a police officer discovered a dead horse on Tchoupitoulas (that's pronounced chop-o-TOOL-is) Street. He immediately found some wharf workers and had them help him drag the horse to Julia Street. A fellow officer saw this operation and asked him why he had done it. The man replied, "I knew I would have to write a report about the dead horse and I had no idea how to spell Tchoupitoulas."

The low opinion of officers' writing, whether justified or not, has prompted agencies to look for ways to simplify it, often by trying to reduce all writing to a few fill-in-the-blanks forms. When this strategy fails, departments find themselves creating more and more forms.

One federal agency developed more than two thousand forms for reporting information, each designed to make life easier for its agents. Needless to say, this plan did nothing to increase the effectiveness of that organization, nor did it make life easier for its investigators. The diversity and complexity of investigations preclude any such simplistic and universal solution.

In addition to disliking writing because of its consequences, many officers view writing as a hindrance to "real police work." They fail to see or refuse to accept that without proper documentation, most investigations have little value. By examining various aspects of law enforcement writing in a systematic way and by presenting writing as an integral part of the investigative process, we hope to remove some of the mystery and diminish the disdain for it. Only with understanding can a change in attitude occur, and attitude is the real problem in police writing.

PURPOSE AND STYLE IN INVESTIGATIVE REPORTING

Using only words, novelists might convey feelings and events with such skill that they earn the highest literary honor in the world, the Nobel Prize for Literature. Scientists, on the other hand, although they too must provide written accounts of their findings, receive awards not for their writing but for the work it describes. By contrasting the two disciplines, we may gain some insight into problems and solutions regarding paperwork for law enforcement.

Examining works of fiction will provide little information for law enforcement to use when developing a model for writing. Creativity and the ability to tell a story in a convincing fashion play little part in effective police writing. In fact, using these talents can lead to disaster.

A scientist's situation, however, offers an excellent example for illustrating an investigator's position. Even though scientists' plaudits result from their discoveries rather than from their reports, unless they articulate those findings to the satisfaction of those who would act upon them, their work will go unrecognized. This coincides with the burden facing today's police officer, who must "discover" the facts and then record them for use by others.

No aspect of police writing creates as much impact as the investigative report, either in terms of publicity or, more importantly, in terms of outcome. Verdicts of guilt or innocence often depend as much on how investigators document their findings as they do on the quality and quantity of evidence obtained. Unfortunately, even in this specialized arena, officers find few rules for recording the results of their investigations that apply to all situations. Therefore, we will examine some of these diverse investigative activities and then suggest some standards or guidelines for documenting them.

THE INTERVIEW

Throughout history, police have tried to gather the facts needed to solve crimes. Just as other researchers would do, they methodically searched for clues and methodically recorded the results. Although made more effective by improved techniques and equipment, today's investigators do many of the same things in much the same way.

Despite the years of practice at gathering physical evidence and the recent advances in forensic science, today as in the past *the interview solves more crimes than all other techniques combined.* People provide the solutions to crimes; they tell investigators "who done it." This very nonmethodical technique remains the primary research tool of most successful investigators.

Almost every investigative agency has at least a few members whose fellow officers describe them as "really good on the street but not worth a damn on paper." These people seem to have the ability to talk with anyone, but their writing, if it exists at all, has little value. Many of these officers spend their entire careers being touted for their expertise while contributing little to the mission of their organizations. An examination of this all-too-common phenomenon may provide some insight for avoiding the stigma of not being "good on paper."

How do these types manage to exist in an environment that relies to such a large extent on paperwork? Often they do so by limiting their efforts to superficial encounters and then moving on, leaving the gathering of details and their subsequent documentation to others. They justify this behavior by defining their mission as one of "greasing the skids." This evasive ploy enables them to survive and in some cases even to flourish. Departments can be well served by using these officers in positions where their ice-breaking skills are paramount. Some of these types make excellent informant developers or liaison officers.

Some investigators really do lack the basic skills and knowledge needed for writing, thus justifying their efforts to avoid any situation in which they must write. However, for most the inability to write has little to do with ineptness. Instead, they fail because they have nothing worth writing despite their willingness to talk with anybody they meet.

Fundamentals of Interviewing

To understand why these bright, likable, enthusiastic people fail so miserably requires us to examine the fundamentals of interviewing; often it is here that their difficulties with writing begin. Their encounters with people ignore the second part of the widely accepted definition of an interview, a conversation with a purpose. They concentrate so much on the conversation that they fail to guide it to the relevant topics, let alone pursue them in depth. Instead they "chat" (some have other, less flattering terms for what they do) extensively about any number of things, declare the interview completed, and depart.

If such officers were to produce a written document based on their efforts, it would leave most readers bewildered. Recognizing this fact, they usually try to avoid writing anything by characterizing the results of their efforts as inconsequential. If this fails, they write either an exceedingly brief account or one that rambles on interminably. The absence of relevant information is the commonality of either effort.

Preparation for the Interview

It is difficult to imagine scientists beginning work on a project without first defining what they hope to achieve and reviewing the work of those who preceded them. So, too, must interviewers consider their purpose for interviewing a person and review all relevant information, such as the facts of the case and the background of the subject to be interviewed prior to beginning that interview. Unlike a scientist, who can usually pause to review previous data, once an investigator has begun the interview, the time for preparation has passed.

Careful planning not only provides interviewers with a clear understanding of their purpose for conducting an interview, but it also helps them to decide *how, when,* and *where* to conduct it, all variables that can contribute to the success or failure of an interview. Unfortunately, investigators who fail to realize this fact, usually those for whom meeting people comes easily, often spend their careers "winging it" and thus rarely obtain the needed information; no amount of writing skill can rectify this. *Failure to obtain the needed information probably accounts for more inadequate reports than all writing inadequacies combined.*

The antithesis of those who "have never met a stranger" are those who fail to grasp the other prong of the interview definition, that *an interview is a conversation.* Their interviews consist of nothing but questions, and they justify this technique on its presumed efficiency. However, treating people as if they are some sort of data bank that will spew forth information on demand rarely succeeds. It did not work when Jack Webb as Sergeant Friday of *Dragnet* popularized it some forty years ago, and it does not work today. Furthermore, a quest for "just the facts" has the added detriment of alienating nearly everyone subjected to it.

An encounter with a representative of the law, no matter how routine it may seem to the officer, is far from routine in the minds of most people, whether they are suspects, victims, or witnesses. This encounter, often the result of some traumatic event, serves to heighten already existing feelings such as fear, guilt, and embarrassment. These as well as many other emotions inhibit a person's ability to provide information. Many investigators choose to ignore these feelings and stick to "just the facts" because it is easier. However, *unless interviewers deal with their subjects' feelings, they will never learn the facts.*

The Dangers of Preprinted Forms
The ultimate technique for avoiding personal involvement during an interview is the use of the preprinted fill-in-the-blanks form. It contains spaces for everything, usually beginning with the person's name and date of birth and proceeding through every conceivable bit of background information about that person. Each blank is usually numbered and the investigator must dutifully start with number 1 and work through the list. If the form is designed for a victim or witness, it will then often move to a series of blanks designed to enable the investigator to obtain a complete description of the culprit or culprits. To make this process logical and systematic, these forms usually start at the top, asking for the suspect's hair color and style and then work their way down the body to the feet. Unfortunately, few people

have thought patterns that conform to the order of the preprinted form. By trying to force recall in this manner, the investigator will stymie the witness's memory and thus fail to get the needed information.

At the end of the fill-in-the-blanks section, these forms usually have a space for the investigator to record the witness's narrative account. By the time the investigator gets to this point, the questionnaire has often conditioned the witness to talk in sound bites. Brevity becomes the standard, and cooperative witnesses comply by severely editing and condensing their information. The investigator will record it, move onto the next interview, and at the end of the day proudly announce the completion of a significant number of interviews.

Despite deadlines and other factors that lend a sense of urgency to a case, *the completeness, accuracy, and clarity of the subsequent report usually determine the outcome of the case.* Failure to obtain the information precludes the possibility of meeting these standards and reinforces the criticism that officers cannot write—they cannot write what they do not know.

Preparing Questions in Advance

Some investigators, aware that no preprinted form can suffice for conducting an effective interview, prepare a list of questions in advance. Although the absence of any visible list of questions can sometimes lend an air of spontaneity to the interview, staunchly adhering to the list often becomes as inhibiting as the preprinted form. Interviewers who compose questions in advance and equate this to preparation miss the point. Dianne Sawyer, noted journalist and television commentator, suggested that interviewers are truly prepared only when they feel free to throw away that preparation. They must be so comfortable with the person and the situation that they have no reservations about where the conversation may take them. Only then can they get the facts they seek.

If preparing a list of questions helps you prepare for the interview, write the list. Indeed, doing so can help you tailor the interview to the person with whom you'll converse and help you make decisions about the how, when, and where of the interview. But be sure not to use the list as a crutch during the interview itself.

Listening

Some years ago, a newly commissioned federal investigator, a former local deputy from the Northeast, received her assignment to an office in the South. Unfortunately, she did not meet the local preconceived notion of a federal agent at that time. She was petite, the "wrong" gender, and had a "strange" accent. As a result, through no fault of her own, few people took her seriously despite her training and experience. Remarks such as, "Hey sweetie, why didn't they send a real agent?" frustrated her on a regular basis. However, at last she located a potential witness who did not question her credibility, a tenant farmer who answered her questions without hesitation. In fact, while standing in a field beside his tractor, he spewed forth such a torrent of information that she had difficulty grasping all of it. As a result, while trying to record this valuable data, she ignored the unintelligible phrase "washemans" that he periodically interjected into his monologue. The farmer seemed to use this phrase, accompanied by a slight nod of the head, in lieu of punctuation. Both the phrase and the gesture were lost on the agent.

To her chagrin, the credibility that she so desperately sought and for once had achieved disappeared when the fire ants began gnawing at her ankles. In his understated way, the farmer had been trying to warn her that she was standing beside an anthill; "watch them ants" was what he had been saying. According to the agent, dignity played no part in dealing with that situation; fire ant bites hurt. She resolved never again to ignore any part of what a person says during an interview. She had learned in one incident what many investigators never grasp—that a successful interview and the resulting written account depend on effective listening. It makes no difference what is asked or how it is answered; only what is heard and understood matters.

Failure to listen effectively during an interview may not always have the immediate reaction that the fire ants triggered, but it can often have even more painful results. Cases can and do go unsolved and criminals go unpunished, not because the needed information was not available, but because the investigator failed to "discover" it, often despite its being provided during an interview. (See Chapter 1 for an exercise designed to improve listening skills.)

Investigators must concentrate on the witness's words and actions, rather than on formulating the next question, allowing their minds to wander, taking detailed notes, or assuming they know what the witness will say. Nothing that goes unheard during an interview will appear in any subsequent writing. Although this is certainly not the result of poor writing skills, it often gets categorized as such; after all, the writing is flawed.

Note Taking during Interviews

Note taking interferes with interviewing. A scientist's laboratory specimens may not object, but taking notes tends to inhibit people while they are being interviewed. It also interferes with the flow of a conversation and prevents effective listening by the interviewer. *However, if investigators hear relevant information that they did not already know and will not otherwise be able to remember, they must record it.* If they do not, it will not appear in any subsequent report.

This apparent conflict may not be nearly the dilemma it seems.

Absent some complicated topic involving numerous facts and figures that require contemporaneous recording, effective interviewers usually *limit* or *completely avoid* note taking during the conversation. After they have thoroughly discussed the topic with the person being interviewed, they move to another necessary phase of the interview, the *verification*. During this phase, while they confirm that they heard and understood what the interviewee said, *they record only the items they will need as prompts for writing the subsequent account.* To record more is counterproductive, and trying to produce a verbatim transcript will completely frustrate the interview process. (In contrast, see Chapter 1 for advice on taking notes from lectures or from reading.)

However, the second stage of note taking, as described in Chapter 1, that of *summarizing* one's notes—in this case, summarizing the interview soon after it has occurred—can be invaluable toward remembering what occurred. (See Chapter 1 for details of this process.)

Having discussed the primary technique for obtaining information here, in Chapter 4 we will deal with some areas that writers should consider before committing their findings to paper.

4

WRITING INVESTIGATIVE REPORTS: PREDRAFTING CONSIDERATIONS

THE IMPORTANCE OF WRITTEN REPORTS

When scientists know for certain that their work has succeeded, they must experience feelings of satisfaction. They must want to shout to the world, "We did it!" However, their work is far from finished. Their methods as well as their results must be reviewed, tested, and scrutinized by many experts before their findings will be accepted. Much of this scrutiny centers on the written account of their research and its results. Knowing this, it is hard to imagine that they would assemble it in a haphazard fashion. With their having devoted so much time and effort to a project, we can presume that their findings will be presented in a meticulous, well-organized fashion designed to make understanding as easy as possible for scientific reviewers.

Police officers, too, must submit a report of their findings, and these findings can be the culmination of many investigative steps. A report may contain written accounts of

- numerous interviews
- surveillances
- searches
- arrests

done by many people; nevertheless, it is ultimately one person, often called the *case officer* or *lead investigator*, who has responsibility for it. Although scientists may have skeptics and critics, the latter might seem supportive compared to some who read police reports. Defense attorneys, for example, hope to discover any errors either in the methodology or the results and to use these flaws to attack the credibility of the investigators. Because of the adversarial nature of this process, officers need to devote the same energy to their writing that they do to solving their cases.

PREDRAFTING/PREPARATION

Just as with interviewing, effective investigative writing depends to a large extent on preparation. An absence of planning becomes obvious to readers, who are often left wondering just what the writing was about. Bewilderment should rarely be the effect a writer seeks. Reports that have this effect lack value; they become reports for their own sake, a far too common situation. To avoid this outcome, writers should consider several items before they begin writing.

Readers and Their Roles

Interviewers cannot afford to concentrate on facts to the exclusion of the human element, and neither can writers. As described in detail in Chapter 2, writers must determine who will read their material and tailor it accordingly. Will an investigator read the material looking for leads to further develop the case? Will supervisors use it to make decisions about devoting additional resources to a case? Will a prosecutor use it to decide if and how to present some evidence in court? Will defense attorneys and their clients have access to the information and try to extract items to impeach the investigator or witness?

Unlike many other situations in life, answers to these questions come easily. However, the answers come only if the writers ask the questions. The solution lies in the asking, and writers who do so and then maintain an awareness of the answers will write more effectively. They will be better able to focus their efforts on a specific purpose and eliminate meaningless and therefore confusing material.

Content: What to Leave In and What to Take Out

Scientists' reports surely omit much of what they did during their research; after all, they may have worked on the project for many years and explored many avenues that lead nowhere. Including all of these elements would overwhelm many readers by sheer volume. Confusion and boredom would beset the rest.

Investigators, too, must decide what to include and what to omit. They might reduce a three-hour interview to a single paragraph; they should if that is all that pertained to the case. While preparing documents, writers should ask and answer the question, "Does this matter?" They should then exclude anything that does not. *This principle does not give writers a license to exclude facts just because the facts do not support their agendas.* Investigators must strive to discover and then record the complete truth regarding an issue. However, much of the information gathered during interviews and other investigative procedures has no relevance to the case and should be omitted.

GETTING STARTED

A blank sheet of paper is an intimidating thing. Clichés about a journey of 1000 miles that begins with a first step and the first sentence of a novel as the hardest to write apply equally to investigative writing. When it comes to writing, many people, including competent investigators who have gathered the relevant facts, have difficulty getting started.

Much of the difficulty of starting comes from the fear of failure, a fear often learned in elementary school. Many adults remember the feeling of terror they experienced as students when they struggled with a writing assignment while their teachers roamed the classroom. These teachers would detect the slightest flaws and correct them, often in a voice that oozed contempt and with sufficient volume for all to hear.

The teachers did not limit their criticisms to content; in fact, they often concentrated on grammar, spelling, and even penmanship. Most people find it difficult to think about facts and ideas and commit them to paper while having to worry about the mechanics of writing at the same time. This preoccupation can produce enough stress that the thought process falters, causing a lack of ideas and the resulting blank page.

Many students ultimately realized that the blank page often prompted less criticism than one that bore any imperfect prose. As a result, as time passed, they became more and more reluctant to begin writing; it was safer not to. Although the teachers have long since gone, the effects of their criticisms remain; the intimidation of the blank page endures.

Outlining and Other Ways to Defeat Writer's Block

Investigators must develop some strategy for overcoming the inhibitions of the blank page. Chapter 1 describes exercises and the regular practice a writing log affords toward working through this difficulty. Another strategy—preparing an outline—also works for many report writers, particularly in instances where a large amount of information contributes to the sense of being overwhelmed that many investigators encounter. It has the appeal of being logical and systematic, a characteristic many law enforcement officers share. Outlining can be especially valuable in dealing with incoherent and seemingly irrelevant information, as often results from interviews. For example, the debriefing of a long-time hoodlum who has reached out to the police because he has fallen from grace with his organization and suspects he has become a target for "retirement" might fit this category. In a case like this, the hoodlum, because of his fears and his efforts to demonstrate his value, may provide a torrent of information, much of which may be only vaguely familiar to the investigator whom he has chosen as his confessor. Likewise, he will probably spew forth his information in a haphazard way, moving from topic to topic and back again with no hint of logic or organization.

Although the initial conversation is hopefully only the first of many debriefings in such a situation, the investigator must document any information initially provided. After all, the guy may be right about his life expectancy; there may be no second chance. This leaves the investigator with the challenge of reducing myriad facts from cryptic notes taken in the same disjointed fashion that the hoodlum provided them. The sheer quantity of the information as well as its lack of organization add to the stigma of the blank page. Where to begin becomes a major concern.

Outlining may solve this problem. By breaking the information into logical categories of manageable size, the chore becomes less daunting. For instance, the investigator might begin by giving the outline an overall title such as "The First Debriefing of Mr. X." Just doing this removes the blank page syndrome. The investigator can then consider the general topics the hoodlum discussed. Perhaps the general topic would be "Criminal activities and the people involved." The writer now has two major subheadings. By looking at them separately, the writer may begin to see some logical way to proceed—such as listing the criminal activities (loan sharking, gambling, and prostitution) and then subdividing these. Under the subhead "Gambling," the investigator might list the various gambling activities to include casinos, sports betting, horse racing; then continue subdividing each until the information has been exhausted. Here is what a portion of the outline might look like:

First Debriefing of Mr. X

    People involved
    Criminal activities
        Loan sharking
        Gambling

Casinos
Sports betting
basketball
football
line information
book makers
identities
locations

By repeating this process for each topic and always leaving space for additional entries, you can ultimately develop a skeleton of the entire session. You can then flesh out the skeleton by converting the words into sentences and the sentences into paragraphs. This systematic approach eliminates the feeling of being overwhelmed and adds some order to what had been chaos.

Freewriting

Some investigative actions, particularly interviews, pose a writing problem not because of the magnitude and complexity of the material but rather because of a lack of such substance. Often an extended interview may consist mostly of rambling conversation interspersed with a few relevant facts. Furnishing the facts in a succinct fashion and still enabling readers to understand what transpired poses problems for many writers.

In such cases, the outline does not offer a solution: the writer does not have enough details to make one. This situation conjures up the same hesitancy as that caused by the elementary school teacher who assigned an essay but refused to provide a topic. This indecision combined with the need to create an error-free document stifles both young students and experienced officers.

Freewriting, a technique that encourages the writer to adopt the attitude that mistakes do not matter, may offer the best solution to this dilemma. Many achieve this attitude by assuring themselves that nobody else will ever see their document, that they are writing it only for their own eyes. Therefore, they need not pay any attention to grammar, spelling, punctuation, or penmanship. The freedom to ignore these long-time inhibitors makes starting much easier.

Others pretend that they are *writing a letter* about the incident to a trusted friend. Trusted friends do not critique letters; they appreciate them. This technique might work well for the First Debriefing of Mr. X, where only vague and sparse information was obtained. People in those circumstances may be so anxious and tense that although they talk incessantly, they provide few substantive facts. Freewriting rather than outlining offers a more effective technique for getting started in such a situation. The result of such an approach might resemble the following:

> Just talked with Mr. X who is scared to death that he might get killed— he wants help and is willing to provide info in return. Will name names—rambled about why they are out to get him. Mentioned various activities including gambling—Mr. X was a successful bookie and went into great detail about the art of bookmaking—insists that being a gentleman of honor is the key—went into detail about his break with the mob—they think he is a snitch. . . .

The investigator could continue on in this manner until memory is exhausted and notes provide no further information. Continuing to write without pausing to evaluate provides the key to successful free writing. Writers must force themselves to ignore the hated but engrained concerns about recording anything in less than flawless fashion.

When they are finished freewriting, writers can review their narrative and highlight those points worth documenting. Then they can extract these items and place them in some logical sequence, perhaps by using the previously discussed outlining process. They can then convert the outline to sentences and paragraphs. For most investigators, this kind of conversion presents little difficulty once the facts are on paper. These ploys and any others that help the writer to begin are acceptable, such as the *dummy drafting* described in Chapter 2.

After overcoming the difficulties of getting started, the writer must now produce the document. Chapter 5 will discuss the actual preparation of an investigative report, including the recording of individual investigative steps and the assembling of these into a finished product.

5

WRITING INVESTIGATIVE REPORTS: FORMAT

SELECTING FROM THE CASE FILE

Nearly every law enforcement agency maintains a separate file for each case under investigation. This case file serves as a repository for all information that pertains to that case. It will not only have the results of both productive and unproductive investigative actions such as interviews and surveillances, but it will also often include administrative material. The file could contain such diverse items as notes taken during an interview, photographs of the crime scene, newspaper clippings about the case, laboratory reports, and a letter from the prosecutor requesting a briefing about the case's progress. Regardless of the complexity, diversity, or magnitude of the file's contents, the case officer bears the responsibility for maintaining and understanding it.

At some point during an investigation, other people such as supervisors or prosecutors may need to know the details of the case. Rarely can the case officer merely hand the file over to the interested party. For some, the irrelevant material would be indistinguishable from the pertinent, and for others, the file might contain material that they have no authority to see. For example, the file might contain information that could reveal the identity of a confidential informant, information that must be limited to those who have a "need to know." The case officer has the responsibility to select the appropriate material from the case file and to disseminate it in report form to the appropriate people.

ARRANGING THE DOCUMENTS

Regardless of the complexity of a case, an investigative report should consist of the file's pertinent documents arranged in a logical order. Report recipients should not have to read irrelevant material, flip-flop back and forth among the pages, nor refer to other documents. *Thus in most cases, the document that initiates the case serves as the first page of the report.* This document, usually prepared by the person who received the information, may have resulted from such diverse sources as a 911 call to a dispatcher, an observation by an officer on patrol, or a discovery by a detective while conducting a search. After selecting this document, the case officer merely adds each relevant document in the logical order, usually *chronological.* However, some cases are so diverse and complex that the case officer must arrange the report in some other way, such as by *types of crimes, locations,* or *suspects.*

Cover Memo and Table of Contents

To complete the report, about the only writing the case officer needs to do is to prepare some type of cover sheet. It may identify the case officer, the suspect(s), the victim(s), and the dates covered by the investigation. Inclusive dates are particularly helpful if subsequent reports become necessary because they enable the reader to place the reports in chronological order. Sometimes cover sheets also contain a one-paragraph summary of the case.

Here is a sample cover sheet for a burglary report:

XYX Police Department

Case Number	Burglary—1234–97
Case Officer	Detective John Smith
Inclusive Dates of the Report	xx/xx/98–xx/xx/99
Primary Suspects	William James Joseph Johnson
Victim	Neighborhood Computers, Inc. 555 Maple Street (Merriwether Plaza)

Summary: On xx/xx/98, the above store was burglarized sometime during the early morning hours. Numerous printers, computers, and other hardware valued at over $30,000 were taken. Suspects apprehended three months later while trying to sell some of the proceeds to an undercover officer in a sting

Continued

Continued

operation. Both suspects have been charged with grand larceny and are in custody awaiting trial. Others are suspected of being involved, and investigation is continuing.

Distribution:
1—Burglary Squad Supervisor
1—District Attorney
1—Case File

Whether or not the investigative report includes other items depends on the magnitude of the case. A report may contain dozens of interviews, numerous surveillances, and other investigative accounts as well as lab reports. In such instances, a *table of contents* may be appropriate.

Witnesses Section

Some cases may have so many potential witnesses that a separate section that *lists* witnesses, their addresses, and phone numbers and provides a *brief synopsis* of the nature of their testimony would prove invaluable to a prosecutor when preparing for trial.

OTHER TIPS FOR WRITING COMPONENT PARTS OF REPORTS

As previously indicated, report writing is a misnomer. Instead of writing a report, the case officer selects the appropriate material and assembles it. Therefore, compiling good reports results from good judgment, not from good writing. The actual writing was done by the participants in the case as they documented their actions and discoveries.

Investigators may not write full reports, but they still must write the individual pages that make up those reports. The critics of inept report writing are actually criticizing these individual accounts. However, merely knowing that the critics have used incorrect terminology does little to remove the anxiety many officers feel. Fortunately, recognizing this error in terminology offers the key to the solution. By viewing writing as a series of manageable steps instead of one overwhelming project, officers can identify and overcome their shortcomings. The following sections identify some of the areas of concern for many police writers and offer some suggestions for dealing with these problems.

Using First Person

Before considering the various techniques involved in documenting the results of an investigation, perhaps we should consider the issue raised most often by police officers in a recent survey conducted by one of us among participants in an e-mail list server discussion group. Asked about writing difficulties in their profession, they universally agreed on a need for clarity, but they could reach no accord as to what that means. Much of the dispute hinged on the question of whether police reports should be written in first or third person. Although the comments implied that many respondents had varying interpretations of those terms, most wondered whether report writers should refer to themselves by name, by title, or by the pronoun *I*. The responses revealed no clear consensus on this issue. Some stressed that using first person contributed to a report's clarity, while others argued in favor of the supposed objectivity afforded by the use of third person. The next section offers an acceptable solution regardless of style preference.

Using Preambles

Some writers manage to satisfy the desire to personalize their writing and still present an aura of detached objectivity by using an introductory paragraph to achieve the former and subsequently employing a journalistic approach to convey the latter. Their introductory paragraph or *preamble* gives the investigators a nearly automatic beginning for recording the results of any investigative action and further reduces the barrier of the blank page. They need only adapt the wording of the preamble to fit the specific type of investigation. Preambles for the investigative activities that make up a finished report can take various forms.

Interview Preamble
A preamble for an interview might read as follows:

> John Doe, of 100 Elm St., Smallville, Michigan, was contacted at his home by Smallville officers William Jones and Mary Smith on month/day/year. After introducing themselves and telling Mr. Doe they were interested in what he had seen at the convenience store the previous morning, Mr. Doe furnished the following information:

This format, or some variation of it, serves many useful functions, and it can be used for almost any interview.

This preamble allows the reader to avoid scanning extensive text to learn the identity of the interviewee; the name appears at the beginning. In a report that contains dozens or even hundreds of interviews, such a tactic can save a great deal of time for the reader who is looking for a specific interview. It also sets forth where, when, and why the person was interviewed, questions that many readers wish to have answered early on.

The preamble also resolves the issue about using first person or third person by rendering it irrelevant. The wording of the preamble identifies the interviewer(s) by name and position, and thus achieves the clarity and personalization desired by some officers. Furthermore, it eliminates any need to make reference to them again. The writer need not say either "Doe told Officers Smith and Jones" or "Doe told us." The reader already knows that, and thus the writers can now merely relate what they learned, much as a newspaper account would do. This strategy lends an air of objectivity to the writing that some officers prefer.

Lastly, the preamble makes it clear that the person interviewed provided all the information that follows it. This makes any further reference to the source of the information unnecessary. The preamble eliminates the need for the writer to use phrases such as "he said" or "he advised," phrases that typically occur throughout many reports and often distract the reader. This format makes them not only unnecessary but also redundant. For instance, at the end of the preamble, the wording "furnished the following information" allows the writer to say, "Doe arrived at the store" not, "Doe said he arrived at the store." The preamble already attributes the information to Mr. Doe.

Search Preamble
Obviously a preamble intended for an interview cannot be copied verbatim for the text of some other investigative activity. However, the same principles do apply. An account of a search might begin:

> Based on a warrant issued on month/day/year by Judge John Brown of the municipal court of Smallville, Michigan, a search was conducted at 100 Elm Street by Officers Mary Smith and John Jones from 2 p.m. to 4 p.m. on month/day/year. As indicated below, the following items were found and retained.

This introductory paragraph sets forth the authority for the search, identifies the searchers, and furnishes the specifics of the search regarding date, time, and location. Like the interview preamble, it allows readers to easily decide whether this document will contain the information they are seeking.

Subsequent paragraphs could then set forth the results of the search using the same criteria as in reporting an interview.

Listing Items Found in a Search.   Qualities such as relevance, objectivity, and clarity apply to all investigative writing, and writers must decide the style and format that best achieves these goals in each instance. For example, an investigator might report the items seized during a search of a gambler's office as a simple list of items found. However, the investigator might record the search of the home of a suspected serial killer in a completely different manner. The potential relevance of the specific location of various items and their proximity to each other might mandate an extensive, step-by-step narrative account of the search and a detailed description of each item seized. Case circumstances would determine the most appropriate format.

Arrest Preamble

On month/day/year, John Doe was arrested without incident at his residence, 100 Elm Street, by Officers John Jones and Mary Smith of the Smallville police department. The arrest was based on a warrant issued on month/day/year by Judge William Brown of the municipal court of Smallville, Michigan. After the arrest, Mr. Doe was immediately taken to the Smallville jail where he was fingerprinted, photographed, and detained pending a bail hearing.

As in the examples of other activities, this introductory paragraph regarding an arrest covers most issues. It not only answers the who, what, where, when, and why questions, it also tells how. It does this with the simple phrase "without incident." These two words enable the writer to omit any details of the tactics used. Giving details about tactics such as locating, handcuffing, and searching a suspect needlessly provides the defense with material to use during cross-examinations.

On the other hand, if the arrest did not occur "without incident," if resistance was encountered, the writer could modify the preamble to immediately alert the reader of this fact. The preamble might read: "After initially struggling with the officers, John Doe was subdued and . . ." Later, the writer could provide the details in the appropriate section of the text. By mentioning the resistance in the preamble, however, the writer ensures that the reader will not overlook it.

Interrogation Preamble

Technically, interrogations are a type of interview. However, their adversarial nature raises some unique issues and mandates some specific requirements. Writers must consider these requirements, such as issuing a Miranda warning, when composing a preamble for an interrogation.

Policies regarding advising people of their rights against self-incrimination vary from one department to another, and many go far beyond what the court mandated. However, all must adhere to the court requirement that people in custody be told of these rights and waive these rights prior to any questioning that could lead to admissions against themselves. By documenting that they did this, officers will not eliminate allegations of misconduct, but they can reduce the number and credibility of such allegations.

> John Doe was contacted at the Smallville jail by Officer Mary Smith on month/day/year. Officer Smith identified herself to Mr. Doe and informed him of the reason for her visit. She then gave Mr. Doe a copy of a form entitled "Waiver of Rights," which he read, stated he understood, and then signed. Mr. Doe then furnished the following information.

Had Mr. Doe refused to sign the form or if Officer Smith had spent time explaining it or even reading it to him, the preamble should also indicate this. The writer should try as succinctly as possible to give the reader a clear understanding of what happened at the outset of the interrogation.

THREE ISSUES: RELEVANCE, ORDER, LISTS

While using preambles can solve some formatting issues in reports, this section notes three other common concerns and offers suggestions for dealing with them.

Relevance

After setting the stage with an introductory paragraph, writers must now present whatever information they possess that relates to the case. They should include little if anything else. As they review their notes, outlines, or freewritten drafts, they should continually ask themselves if a point matters to the reader. Doing so will help them avoid cluttering their writing with extraneous information and will enhance its clarity.

Order of Information

Writers must also decide the proper order in which to present their material. Investigative activities have a beginning, middle, and an end; so does the information they produce. As a result, a *chronological* accounting of an event usually works best. This ordering rarely presents a problem in recording the results of interviews, inasmuch as witnesses and victims usually provide accounts of events as they occurred. However, sometimes they do not. For instance, a cooperative suspect may begin his story by discussing his recent arrest. He may then talk about his associates and move from there to how he initially got involved. Even so, if interviewers take the time to verify and understand what they heard, they can usually rearrange these bits of information into some orderly fashion, chronological or otherwise. The previously discussed *outlining* technique can often facilitate this organization process. *Investigators should feel no obligation to report items as they received them. Instead, they should always strive for clarity for the reader, and presenting items in a logical order can help achieve this goal.*

Handling Lists of Information

Many writing manuals stress the importance of using clear, concise, grammatically correct sentences. The preamble and body of an account of an investigative effort should usually meet this criterion. Because of the need for clarity, however, not all documentation of investigations lends itself to sentences and paragraphs. For instance, if Mr. Doe had given a detailed description of an associate during his interview, putting that data in paragraph form would look something like this:

His associate is a white male, about forty years old, nearly six feet two inches tall who weighs approximately 230 pounds. He has dark-brown hair that is short on top and long on the sides. He has brown eyes, a fair complexion, a heavy mustache, a one-inch circular scar just below his right eye and a tattoo of an eagle on his right shoulder.

This description requires more than a cursory glance by readers to know what the person looked like; they would probably have to read it several times. Contrast this description with the following:

Race	White
Sex	Male
Age	40 (approximately)
Height	6'2"
Weight	230 lbs.
Hair	Dark brown (short on top, long on sides)
Eyes	Brown
Complexion	Fair
Scars	One-inch, circular below left eye
Tattoos	Eagle on right shoulder
Other	Heavy moustache

By putting the description in column format, the writer makes it much easier for the reader to envision the man. The more detailed the description and the greater the number of such items, the more important this clarity becomes. Imagine reading a paragraph that gives the results of a spot check of a parking lot, one that contained dozens of cars. Looking at that many cars of various makes, models, years, colors, and license plates in paragraph form would overwhelm most readers. More important, imagine the difficulty this kind of description would cause for someone charged with entering the data into a computer.

Although setting out descriptive data in columns aids readability, inserting these columns within the body of the narrative detracts from the overall appearance of a document. Writers can avoid this problem by putting all such descriptions at the end of the document. This method requires only a brief explanation such as:

> Descriptions obtained during the interview, including that of Mr. Doe, are set forth below.

OBTAINING FEEDBACK ON DRAFTS

When police officers tell "war stories," particularly traumatic ones, about their work, their partners almost always play a significant role. Officers develop extremely close relationships with their partners, often closer in many ways than with their own families. They have few if any secrets from each other. Yet few officers ever ask their partners to review their writing even if it recounts an event in which they worked together.

After having gone through the process of preparing a document, few people, including investigators, want somebody else to read it, much less change it. After all, they are the ones who did the investigation, and it should be reported in their style. Besides, they have read it several times and have even run it through spell-check; that should suffice.

Unfortunately, the adage about attorneys who defend themselves having fools for clients also applies to writers who edit their own work. Writers tend to read what they meant, rather than what they wrote. Writers need somebody else to find their mistakes, but ego or a reluctance to impose often prevents them from asking for help. Given the opportunity, defense attorneys will gleefully fill this role, but few officers relish this kind of "help." Defense attorneys will use the writers' mistakes to attack their credibility, offering the errors as evidence of incompetence and dishonesty.

To avoid this outcome, officers need to take advantage of the relationships they have developed with their partners—to trust others with your life but not trust them to critique your writing seems absurd. Besides, having one's mistakes corrected by a trusted friend in no way approaches the discomfort experienced when being cross-examined on the witness stand.

Furthermore, by offering to reciprocate, officers can remove the stigma of imposition. They can also make the review process easier *by asking their reviewer specific questions* about the document and by noting any areas of concern, as described in detail in Chapter 2. This technique not only helps reviewers focus their attention, it encourages them to provide concrete suggestions rather than generalized observations. Typical questions could include:

1. Which parts did you have to read more than once?
2. Where are there gaps in the information or any conflicting statements?
3. What in the report might a defense attorney readily misinterpret?
4. Would you be comfortable testifying based on this document, or would you change it?

Even when they are carefully written and logically assembled, reports can still contain stumbling blocks that detract from their effectiveness. Chapter 6 will deal with some of these pitfalls.

6

WRITING INVESTIGATIVE
REPORTS: PITFALLS

College Greek societies subject their prospective members to initiations
and secret rites that foster both loyalty and a sense of uniqueness. Often
the worse the ordeal, the stronger the allegiance. Law enforcement personnel also belong to a tightly knit community, one that occasionally subjects
them to life-threatening ordeals. Members of this society often acquire their
own vocabulary, codes of conduct, and prejudices. Sometimes these unique
characteristics find their way into writing intended for readers outside
the group. Officers need to be alert to this tendency and guard against
several manifestations that would reflect poorly on the writers or their
departments.

POLICESPEAK

"The alleged male perpetrator proceeded to exit said vehicle." The e-mail
survey of law enforcement officers from around the world noted in Chapter 5 suggested that many people view a sentence like this as the epitome of
what is wrong with much of the writing done by police officers. Although
most officers know that a better version would be, "The man got out of the
car," they disregard this knowledge as soon as their writing becomes official business. Much police writing continues to resemble an oldtime movie
script written for Leo Gorcey, the leader of the Bowery Boys. His use of
such phrases as "aforementioned prestidigitations" baffled everybody and
enlightened no one. The scriptwriters did this to amuse the audience—police reports have no such purpose.

The tendency to write in *policespeak* occurs throughout law enforcement. Neither size or type of department nor officers' experience level seems to inhibit it. Many officers attribute this convoluted writing style to the nature of their opposition—defense attorneys. In an effort to "outlawyer the lawyers" by sounding sophisticated and judicial, officers sacrifice clarity and conciseness. Defense attorneys take advantage of this flaw and use it to attack the officers' credibility, the very attribute they were trying to emphasize.

Regardless of the cause of the poor prose, officers must overcome this misguided inclination for creating confusion, not an easy task. Having a partner review the draft rarely works; after all, the partner belongs to the same society. Even the department's clerical staff become conditioned to this style of writing and will thus fail to correct it. One technique that often helps depends on the writers' willingness to look at each sentence and ask themselves if there is a simpler and clearer way to write it. Some successful writers view each sentence as if it were intended for their twelve-year-old son or daughter. Just asking this question provides the answer: "Would I ask my daughter to *secure the rear exit to our residence* or would I would I ask her to *lock the back door?*" The key to success is in the asking.

TACTICAL TALK

Another problem comes from officers' desire to sound official in their writing. When involved in "operational situations" (i.e., real police work), officers value efficiency in both words and actions. Prompt, decisive actions provide a margin of safety, and succinct radio messages help make them possible. Officers do not want to hear long rambling radio transmissions when a few words would do.

Those who try to duplicate this operational style when writing can produce documents that read like the dialogue from a grade-B cave-dweller movie: "Saw suspect in house, arrested him." Eliminating subjects, verbs, adjectives, and adverbs from a written narrative of an event does not convey a professional image. Instead, it gives the impression that the writer thinks that a capital letter followed by a period equals a sentence.

In some instances, the desire for brevity stems from defensiveness. Investigators, particularly those who have been brutally cross-examined, often adopt the philosophy that the fewer words they write, the less they will have to defend. Unfortunately, good defense attorneys can also detect omissions and may use them to attack the writer's integrity by implying some type of coverup. Regardless of investigators' intentions, an overzealous effort to minimize their writing can result in the omission of facts as well as words, facts needed by the prosecutor or other readers for whom the account was written.

WORDINESS

Just as some officers attempt to convey professionalism by being too brief, others take the opposite approach. They assume, as do members of many other professions, that a document's importance depends on its volume and the size of its words; in each case, the bigger the better. Writers who use this philosophy produce vague, convoluted documents that defy understanding. Bewildering a reader rarely enhances a writer's stature.

Investigators can reduce this bewilderment by viewing their documents as crime scenes that the reader must examine for evidence. Just as irrelevant items clutter a crime scene, vague or excessive words clutter a document. In either case, the more clutter, the more difficulty. Before Lincoln gave his Gettysburg address, he was preceded by a famous orator who spoke for two hours. How many people now know who that person was or what he said? Good writers, like good speakers, use only the words needed to convey their ideas.

JARGON

Any outsider who has ever listened to a technocrat can understand why research companies hire writers to describe their products. Without these writers, the company's products would go unsold because many of the researchers lack the ability to articulate their work in lay terms. Police officers also face this problem. Much of their language, particularly acronyms and abbreviations that they take for granted, has little meaning to outsiders. "The decision was made at SOG, rather than by the SAC": The writer of this sentence intended to tell the reader that responsibility for the decision lay with headquarters (seat of government, or SOG) rather than a field office (special agent in charge, or SAC). Even though the writer understood this, few people outside the organization would have.

Even worse, much police terminology has multiple meanings, thus forcing readers to puzzle out an interpretation. "The unit remained operational" could refer to a tactical team's readiness, or it might mean that the officer's car was still drivable. "We secured the area" could mean they established a perimeter, or it might mean that they departed the scene. The reader must put these sentences into context to determine their meaning. Use of such language increases the chance of misunderstanding.

PREJUDICES

The process of dealing with criminals on a daily basis makes cynics of most officers. Sooner or later they come to regard everyone as a criminal—all are guilty until proven innocent, and even then they are probably guilty of something. This attitude hinders investigators by destroying their objectivity in dealing with people and in evaluating the information they collect. Then they reveal this cynicism in their writing, and it hurts their credibility.

If most of their work involves specific ethnic, racial, or other cultural groups, this cynicism can also lead to the acceptance of stereotypes. Indicators, some subtle and some blatant, of these prejudices can creep into investigators' writings, and defense attorneys will use them to divert the court's attention from the issue of guilt or innocence. Instead, the officer's character and suitability become the issue. Defense attorneys have gotten more than one acquittal using this ploy.

Police cynicism does not stop with the general public. They often perceive that judges, politicians, and their own bosses place unreasonable restrictions on them while demanding ever-expanding services. These perceptions, justified or not, affect the actions of some officers. They sometimes respond with acts of malicious obedience that can cause problems for them. A police officer who drags a traffic violator from his car and forces him to a spreadeagle position, a procedure clearly intended for dangerous criminals, will get no more support than a coach who demands that girls on an adolescent sports team obey the rule requiring players to wear jock straps.

The same principle holds true for police writing. Not just blatant terminology but any language designed to ridicule a portion of society has no place in a police report. No matter how harmless or amusing it may seem to some, its presence hurts the image of the writers and their profession. For example, officers who refer to manhole covers as personhole covers leave no doubt about their attitudes on a mandate to avoid sexist language.

Although most officers do not deliberately try to offend others with their writing, they sometimes do so inadvertently. They can be oblivious to the offensive nature of words and phrases. On other occasions, writers offend through a misguided attempt at political correctness. Although use of the term *policewoman* lacks the sarcastic intent of *personhole cover*, it still distinguishes an officer by gender. Failing a valid reason for making such distinctions, writers should try to avoid them, and they usually can. In this instance the term *police officer* works quite well.

Unfortunately, writers have difficulty detecting these kinds of problems even when they look for them. Because *they* do not feel offended, they fail to understand or even notice that their words might insult others. To overcome this discrepancy, they must adopt the attitude of the public speaker, who, faced with the question of whether or not to use some material, should already know the answer: if you need to question material, do not use it. Not only will this tactic eliminate offensive words, it will help writers to select more precise terms and thus enhance the quality of their writing.

ASSUMPTIONS

Officers often make assumptions during their investigations. If they see *A* and *C*, they correctly assume that *B* must also have occurred. However, during a fast-moving, tension-filled situation, the difference between an observation and an assumption can become blurred. This failure to distinguish between the two can cause problems.

During a surveillance, an agent watched two people, one of whom was carrying an envelope, enter a restaurant. When the couple left the restaurant, the agent saw the other person carrying the envelope. The agent's account indicated that one individual had given the envelope to the other. However, under cross-examination the agent admitted that he had not seen the envelope change hands. Following this admission, the defense forced him to acknowledge that he had recorded an assumption as a fact. The defense then asked, "What other facts in this report are really assumptions?" followed by, "Is there anything in this report that is not an assumption? Can we believe any of it?" Nobody who has experienced this type of treatment ever wants to endure it again.

To solve cases, investigators must combine their observations and experiences to reach conclusions and make decisions. However, in reporting their results they must take care to avoid confusing facts with suppositions. "Do I know this happened or do I think it happened?" is the question writers must continually ask themselves. Maintaining an awareness of the tendency to commit such mistakes is often sufficient to prevent them from occurring, or at least it helps one detect and correct them.

JUDGMENTS

Readers make judgments based on the facts provided by the investigators. Investigators must not provide the conclusions. Officers who have worked road patrol know this fact well. Regardless of how drunk a driver might appear, they refrain from classifying the driver as drunk. Instead, they describe the symptoms that led them to that conclusion. Their writing will show that the driver had slurred speech, emitted a strong odor of alcohol, and could not walk a straight line. The readers will conclude that the driver was drunk.

One situation in which officers have difficulty omitting their opinions occurs when they conclude that a person has lied to them. The feel compelled to tell their readers, particularly other officers, of their opinion because they know that merely relating what they heard will be misleading. Rather than inadvertently reporting an assumption as a fact, they deliberately include an opinion despite awareness of the possible consequences.

They could avoid this dilemma by rejecting the fallacy that interviews consist solely of words. The nervous mannerisms, a lack of eye contact, and whatever else they saw all contributed to the interviewer's knowledge; they are part of the interview process. Had they reported these observations, the reader could have reached the same conclusion they did—the person lied.

(See Chapter 1 for exercises intended to improve observation. The Taking Notes section in that chapter mentions the crucial role that body language plays in communication and understanding and recommends that such observations be included in notes.)

MISUSE OF QUOTES

Using the exact words of a recognized figure in giving a speech or writing an essay can add impact and credibility. In an investigative report, a criminal's exact words may occasionally reveal a uniqueness that could prove valuable later on. Lacking this type of situation, writers should use quotes sparingly, if at all, and they should never use them to disguise investigative shortcomings.

To Conceal Vagueness

If interviewers hear ambiguous words or phrases during an interview, they must clarify them. Failure to do so will present problems when they try to reduce the interview to writing. A bank teller may say that she "buzzed out" during a robbery. Quoting this phrase will not help the reader to understand it. Did the teller faint, make a phone call, or escape using an exit that had a hidden electronic lock? The reader cannot determine the distinction, but the interviewer could have. Quotation marks will not salvage poor verification during an interview.

To Justify Obscenities

Even though criminals often direct obscene language at their victims, investigators rarely need to repeat these obscenities in their written documents. Some investigators justify doing so on the basis that it helps to convey the callousness of the criminals and their deeds. Others include them for their shock value. The loss of professionalism and respect caused by the presence of obscenities in the investigative report more than offsets any value they provide. Putting the obscenities in quotes does not negate this negative effect.

An investigative report consists of a compilation of written accounts of the various activities performed by those involved in the case. If the investigators effectively gathered the information and properly recorded it, assembling the report poses few problems. Assembly rarely involves more than arranging the separate accounts in a logical order and preparing a cover page. This principle holds true regardless of the complexity of the case. However, if the investigators performed poorly either in the collection of the data or in recording it, no amount of shuffling of the pages will remedy these flaws.

7

WRITING MEMOS

THE NEED FOR STRONG
INTERNAL COMMUNICATION

"I know you need more help; you and everybody else on this squad. Put your request on paper and I'll consider it along with all the others." Law enforcement paperwork goes far beyond recording investigations. Officers must document actions, provide explanations, and make requests of many kinds. The old adage "If it's not on paper, it doesn't count" still holds true. Even in the age of the computer, which proponents said would reduce reliance on paper, the number and size of documents—paper as well as electronic—have dramatically increased. Stated goals notwithstanding, the greater the ease of communication, the more documents organizations tend to produce—especially internal documents.

Even the effort to reduce an organization's reliance on memos and reports can lead to more paperwork. Consider the following anecdote: At an all-department meeting, executives of a large investigative agency stressed the need to reduce that agency's paperwork. They suggested that eliminating needless duplication of documentation would make a significant contribution to this goal. As an example, they said that if they recorded information furnished to them by an employee, the employee should not document the same information. They also recommended that employees reduce the size of documents by restricting them to the issue at hand rather than providing a lot of needless information; if they asked what time it was, they said, they did not want to know how to build a watch.

Then the executives solicited ideas about addressing these problems. Somebody suggested that the training department teach all new employees how to write concisely, a skill few had. A training department representative disputed this contention. He pointed out that although he did not know the exact numbers, he was certain that he could get the data showing that his division had tested many trainees and that few had significant writing deficiencies. The executive asked the trainer to get the test results to him as soon as possible.

The following day the trainer phoned the executive and said that the training department had tested 1,112 trainees and that only six had shown serious writing deficiencies. The executive replied, "Can you put that on paper and send it to me?" The trainer prepared a memo setting forth the results of the testing and forwarded it to his boss for approval. The boss returned it with the notation, "Beef this up a bit before we send it to headquarters. Tell them about some of the great stuff we are doing here with our trainees."

As Pogo said, "We have met the enemy, and they are us."

MEMO FORMAT

Chapters 1 and 2 deal with many of the techniques and concepts that contribute to successful memo writing. Predrafting, outlining, drafting, and revising can often improve writing regardless of the document or the profession. Clarity, conciseness, and grammatical accuracy also contribute to the quality of any document. Although those chapters do not deal with memo writing per se, the advice offered there is readily adaptable.

Formats for internal documents vary greatly from one agency to another; nevertheless, many have common features that merit consideration here.

Headings

Horror stories abound about documents gone unread until too late, either because they were misdirected or because they were in an inbox where they gradually sank to the bottom. Although nothing can guarantee this fate will not happen to any given memo, the writer can reduce the chance of either of these outcomes by *making the recipient and topic noticeable* at a glance in the *heading.*

Unless the agency uses a preprinted form or, more likely, a template on a computer, the writer should

1. *Date* the memo.
2. Using official names and titles, *list* the recipient at the top, followed by the *sender* and the *topic.*

Writers have little if any leeway regarding the first two categories, but how they word the topic or title may determine the attention the memo will get. Consider the following memo titles: "Need for Vigilance" versus "Enemy Attack Expected." Although the contents might be the same, the latter title will probably get a quicker response. Headings should never merely announce a general topic; they should *alert the reader to the type of action needed.*

References

Writers often produce memos whose effectiveness depends on information contained in other documents, which sometimes require furnishing copies of those documents or restating their information. If writers know that their recipients have access to the other information, however, they can merely *call attention* to it. Most readers would prefer reading one line on a memo— "Reference my report of October 18"—to receiving an attached report of sixty pages, a report they already had received.

Getting to the Point

Despite administrators' constant demand for full documentation, few want to spend much of their time reading the material they insisted the officers produce. As prime minister of Great Britain during World War II, Winston Churchill reputedly insisted that his advisors limit their memos to one page. Although writers deal with many topics they cannot cover adequately in one page, they should strive for brevity whenever possible.

Many administrators regard themselves as action-oriented people who want to get to the bottom line; they often say they "skimmed the material." Few have taken speedreading courses, so what they really mean is that they read the first few sentences and then skipped to the bottom of the page. Writers must take advantage of this practice by *immediately stating their main point or premise* and following this statement with supporting data. If they have a specific *request* based on the data, they should put it either at the beginning—if the request is the main point of the memo—or at the end, the other likely place that a busy reader will look. *Never bury the request in the middle of a page or document.*

Such careful placement improves the chances of the request's being granted or at least considered, simply because it is more apt to be read.

Copies

If writers send copies of a memo to several recipients, they should note this fact on the memo. Sometimes just knowing that somebody else—especially someone whose good will your recipient wants to keep—has received a copy can prompt action that might otherwise not happen. *Caution:* While sending a copy of your memo to your recipient's boss may prompt your recipient's action, judge carefully the risk of needlessly antagonizing your recipient before sending copies up the office hierarchy.

Moreover, writers should *always make a second copy* for themselves. In the age of computers, this copying can take the form of an electronic copy instead of a printed page. However, most computer experts recommend that the copy be stored on a separate disk rather than, or as well as, on the computer itself. One failed hard drive can obliterate years of records.

Sample Memo

Although many formats can fulfill a memo's requirements, a generic memo based on the points we have discussed might resemble the following:

Date: xx/xx/xx

TO: Sgt. Marie Jones, Fraud Squad II
FROM: Det. William Smith
TOPIC: Overtime for surveillance team
REF: Position Paper of Chief Jakob dated xx/xx/xx

BODY: The surveillance team became entitled to overtime pay as of two weeks ago. As mandated in the referenced paper, each member of the team has worked the required number of hours for each of the last four pay periods. These hours were devoted exclusively to surveillances requested by Fraud Squad II. It is requested that you execute the necessary forms for overtime to be paid to the surveillance squad.

COPIES: 1—Chief Jakob, 1—Detective Smith, 6—Surveillance Squad

---

PURPOSES/TYPES OF MEMOS

Writers must consider the purpose of their memos and tailor the contents accordingly. (See Chapter 2 for tips on clarifying the purposes of any piece of writing.)

Depending on their purposes, memos dealing with the same topic can vary substantially. The following section deals with three typical types of memos that correspond to differing purposes. We include a short sample of each type.

Memos to Document

Officers prepare some memos with the absolute certainty that nobody will ever read them. *Memos to document* are often regarded in this manner, because they merely provide supporting data or evidence of action taken rather than initiating new action or making a request. For example, such a memo may result from a requirement that officers notify their organization that they attended some mandatory training program. Officers know that a clerk will glance at the title, check an appropriate block on some form, and file the memo where nobody will ever see it again. As a result, they pay almost no attention to the contents when they write it.

Sooner or later, disaster results. Three years after the writing of the innocuous training memo, a lawsuit makes it a big issue. Dates of the course and topics covered, along with numerous other things that the officer could have documented in a few sentences, become critical. Unfortunately the memo does not contain that information, and its writer learns the hard way that "if it ain't on paper, it didn't happen."

Sample Memo
A chief asks the head of the burglary squad to give him an account of the current situation. The memo text might run as follows:

Sample Memo to Document

TO: Chief McMillan
FROM: Lieutenant Suarez, Burglary Squad
TOPIC: Current Situation
DATE: xx/xx/xx

The burglary situation in this precinct is not encouraging. The number of burglaries reported during the past two years has nearly doubled from the previous two-year period, but the number of solutions has remained constant. Of the 315 reported burglaries, over 200 were of residences, 75 were of businesses, and the rest were distributed among schools, churches, and automobiles.

Copies: 1—Lieutenant Suarez

Despite the brevity of this text, it achieves the intended purpose: providing the reader with an understanding of the nature and scope of the problem. A one-sentence memo, "Things are bad and getting worse," although accurate, would not. On the other hand, the writer avoids cluttering the document with irrelevant data. The memo contains no details regarding specific cases, investigative incentives, or other information the chief had not requested.

Memos to Explain

Because everybody makes mistakes, sooner or later all officers find themselves in the uncomfortable position of having to explain some transgression. Many such incidents have no real consequences and require *memos of explanation* only because they happen to be a pet peeve of the boss. Much of what many in law enforcement hate about paperwork falls into this or some similar category: what they call "paper for its own sake." Officers must beware of treating these memos of explanation in too cavalier a manner or, worse, using them as a means of ridiculing authority. For instance, a chief may regard the department's weight standards as critical, and when officers weigh in over the limit they must write memos of explanation on what they are doing to correct the problem. One officer might respond that because the weight limit is based on a height-to-weight ratio, he is not trying to lose weight but is instead doing everything in his power to grow taller. What could have been a three-sentence explanation becomes an ongoing battle with the boss. No good ever comes of such contests.

Because of the perceived triviality of many required memos, some officers tend to regard all of them, particularly those requiring explanations, with contempt. This contempt leads them to write superficially to meet the requirement for submission while providing little real information. Years later, when the incident results in an administrative or legal action, writers have none of the details they need to defend their actions. Now they must write memos to explain the lack of earlier explanation, but these belated efforts lack credibility.

Sample Memo

A memo in response to a chief's question, "What are we doing about the burglary situation?" would differ from the previous request for a status report. Again, although a one-line response such as "We are devoting considerable resources to the problem" might be accurate, it would not likely satisfy the chief. A better option might read:

Sample Memo to Explain

TO: Chief McMillan
FROM: Lieutenant Suarez, Burglary Squad
TOPIC: Actions to Respond to Increase in Burglaries
DATE: xx/xx/xx

As a result of the increase in burglaries, we have doubled the number of officers assigned to work these cases. Because the largest increase is in house burglaries, patrols have been instructed to devote a greater percentage of their time to residential areas. In addition, detectives have been told to increase their efforts to develop informants in these areas. We have also been in touch with several insurance companies who have agreed to provide significant cash rewards to anyone who assists in recovering stolen goods.

COPIES: 1—Lieutenant Suarez

Memos to Request or Propose

As indicated earlier, organizations function because of paper, much of it internal. Not just administrators but also street-level officers, if they are to be successful, must write effective *memos to request;* that is, proposals. Whether or not they get the needed assistance in a complex case, obtain permission to change shifts, or receive authorization to attend an in-service training program may depend on how well they write their requests. Using some of the techniques described in the following paragraphs may increase a writer's approval rate.

Memos to Persuade

Some law enforcement personnel consistently get suspects to admit willingly to otherwise unprovable crimes. Officers who are able to do this understand the art of persuasion. Yet many of these same officers, when confronted with the need to write a memo designed to convince others to do their bidding or honor their requests, fail miserably. They provide suspects with good reasons to confess but do not give administrators good reasons to approve their proposals. They abandon the very skills that make them good interrogators.

Successful *memos to persuade* often result from the same persuasive techniques interrogators use, and examining some of these tactics can help enhance persuasive writing. The intent of this section is not so much to teach new skills as to make you aware that the skills you already use in one arena can apply to others. Officers who can sell suspects on the idea of going to jail should certainly be able to sell an administrator on the approval of any reasonable request.

Interrogators do not get confessions by asking, "Did you rob the store?" Instead, they try to offer the suspect logical or emotional reasons to confess. One such ploy, *creating a sense of urgency,* can sometimes prompt an admission. In the case of the robbery, they might suggest a need to react quickly: "You were not alone in this, and we got two of your partners. The first one who talks might get a break; after that, cooperation will mean nothing. Who's going to get some consideration, you or one of those other guys?"

This same sense of urgency can often be applied in a memo. For instance, in requesting resources for a sting operation, the writer might suggest, "Experience shows that the peak season for burglaries is in the next three months. If we are to take advantage of this fact, we need to begin our operation at once, something we cannot do until the resources have been allocated. The results we can get with the same efforts three months from now will pale in comparison to what we could achieve right now."

Interrogators sometimes succeed because they manage to create the impression that they are entitled to a confession, that *the suspect owes it to them* in return for something they have done for them. "I believe you when you tell me you did not get any money from the robbery. You know why? Because this wasn't your idea—you just went along for the ride, and I am going to make sure the prosecutor knows that; I am not going to let him blame you for planning this thing. You were just along for the ride, weren't you?" By accepting the suspect's version of the crime and offering to defend it to others, the officer has put the suspect in his debt.

By the same token, the memo writer might note, "During the past six months our squad has conducted more investigations and cleared more cases than at any time in the past. This performance earned plaudits from the front office for the entire squad. However, the squad now needs some relief, or the quality of their work will begin to suffer." Supervisors who read such comments may well recognize that their success has resulted from the efforts of others and feel indebted to them.

Good interrogators *use the suspect's own words* to compel a confession. "You just told me you were at the scene. Now I am telling you we can show that there was only one person at the scene, the one who did it. You also told me you were a man who accepts responsibility for his actions. Based on what you said, that you were there and that you are a responsible man, then you must admit that you did it."

Similarly, the memo to the supervisor might say, "As you indicated in your all-personnel directive of six months ago, 'Anyone who can get the cases will get the resources needed to work them.' Since then, my partner and I have opened far more cases than needed to meet your directive. Based on your statement, we are requesting the assistance needed to work them." This approach has a much better chance of success than a memo whose wording amounts to, "Give us the help you promised."

ADOPTING A POINT OF VIEW

Interrogators, sales representatives, and writers of memos have one thing in common: they lack the authority to demand compliance. Absent this power, they are left with only the opportunity to persuade; successful ones do it effectively. They achieve this effectiveness not just by using various persuasive techniques but also by tailoring their presentations to the individual who can grant their requests. (See Chapter 2 for tips on clarifying your sense of the audience in any writing situation.)

Unlike investigative report writers, who must write in an objective manner devoid of opinion (as described in Chapter 6, under Judgment), memo writers often have no such restriction. They strive to convey their opinions, to convince others of the correctness of their views, and to influence the behavior of their readers. *Memo writers have a perspective* and often have no reason to convey opposing views with the same zeal that they do their own.

Although writers of both investigative results and memos must consider their audiences, the former have to deal with a variety of readers with different backgrounds and agendas, while the latter usually have a more select readership, often only one person. This narrow focus offers memo writers an opportunity to tailor their writing to that person's style and taste. Just as investigators should learn as much about their subjects as they can before interviewing them, memo writers should invest the time needed to understand the recipients of their efforts.

(See Chapter 2, under Audience, for procedures and questions to use in tailoring your writing to specific readers.)

Readers and Personality Type

Although we must avoid the temptation to stereotype any profession or role, realizing that people tend to gravitate to careers that suit their personalities can sometimes help in dealing with them. The following section compares just one aspect of two different personality types and their tendency to assume different roles within law enforcement. Perhaps this example will illustrate how a writer's awareness of the reader's perspective, value system, and style can contribute to successful memo writing.

Understanding Your Own Personality

Writers need first to recognize their own perspectives and values before trying to understand those of other people. Although every person is unique, a variety of instruments designed to quantify various aspects of personality reveals an amazing number of similarities among most law enforcement personnel. One such instrument, designed by David Keirsey and Marilyn Bates, classifies people into four categories or temperaments, which they label "guardians," "idealists," "rationals," and "artisans."* When the Kiersey-Bates survey was administered to over 1,000 law enforcement personnel, the results placed well over 90 percent of them into one group, the "guardians," a category that makes up only 38 percent of the general population.

Guardians display a mix of characteristics that distinguishes them from the other temperaments. Other temperaments lack some of these traits or possess them to a lesser degree. These characteristics include:

1. A reliance on senses rather than on intuition
2. A respect for traditions
3. A desire for clearly defined parameters of behavior
4. A philosophy expressed by the motto "If it ain't broke, don't fix it"

However, the dominant characteristic of the guardians is their *sense of duty,* their need to serve. Knowing that something needs to be done provides them with sufficient motivation to act. The degree to which they feel this obligation to contribute sets them apart from all others.

Successful investigators, most of whom belong to the guardian type, realize that these values have little impact on criminals. Respect for rules and duty has little appeal for most confirmed lawbreakers. Although criminals may give these values lip service when it benefits them to do so, the values of the personality group Kiersey and Bates label "artisans," such as excitement, adventure, and a freedom from restrictions, have more appeal for lawbreakers. Ineffective investigators ignore this difference and instead often berate criminals with the refrain, "It's the right thing to do." This tactic rarely works regardless of how often or how loudly investigators say it. Instead, the statement "Although it might be a bit dangerous, there is good money involved, and it could provide a real rush" offers a better chance of success. Whether developing an informant or seeking a confession, good investigators appeal to the criminals' values rather than their own.

---

*David Kiersey and Marilyn Bates, *Please Understand Me* (Del Mar, CA: Prometheus Nemesis, 1984).

The Personalities of Proposal Readers

Unfortunately, many effective investigators who know that their values hold little appeal to criminals fail to realize that their value systems may not always appeal to their fellow officers, either. Members of the law enforcement community represent various personality profiles, including a category Kiersey and Bates refer to as "rationals." Unlike the guardians, rationals rely on intuition rather than on their senses, and they are more motivated by a desire for improvement than by a need to serve. Although they make up about 12 percent of the general population, experience shows that the number of rationals in law enforcement is far fewer.

Though few in number, the rationals should not be ignored by the rest of law enforcement: the rationals' desire for improvement combined with the intuitive thinking that enables them to envision their organization's futures makes them prime candidates for promotion; rationals tend to achieve leadership roles. An informal survey of the membership of a state police agency showed that although the department had only a few rationals, they headed thirteen of the department's sixteen divisions. Thus, the memos written by guardians are often destined to be read by rationals.

Although rationals may appreciate the concepts of duty and a need to serve that the guardians value so highly, they are far less motivated by these ideals. Instead, rationals tend to look for the big picture and the bottom line. "What will be the total impact of my approval or rejection of this request?" may well be the question that the administrator asks before making a decision. Writers who have addressed this question have a better chance of gaining approval.

Although a reader's perspective may be a minor concern when writing memos to document or explain, it becomes paramount in writing proposals or requests. Consider the following versions of a memo designed to obtain additional resources:

Version 1

The number of burglaries in this precinct has increased dramatically in the past two years. Citizens are afraid to leave their homes for fear that nothing they own will be there when they return. As the burglaries have increased, so have the insurance rates, and most people can no longer afford adequate coverage. Our citizens have a right to expect to feel free to come and go as they wish, and it is our duty to ensure that they can. By setting up a storefront to serve as a fence for stolen goods, we could identify most of the major players and get them off the street. The resources needed to meet this responsibility are . . .

Version 2

During the past two years, despite doubling the number of officers on
patrol, house burglaries have tripled while cases cleared have remained
stagnate. Meanwhile, many higher-profile cases that may determine
how headquarters will allocate support remain inactive. Establishing a
storefront operation would enable us in three months to make more arrests and recover more merchandise than our precinct has done in the
past two years. This improvement would reduce the burglary problem
to a level that would enable us to reassign officers to the priority programs. The resources needed to solve this problem are . . .

Depending on the reader's perspective, either the first memo, which
stresses obligation, or the second, which emphasizes outcome, has a better
chance of approval. The advantage does not depend on the quality of the
request but on the reader's outlook. Administrators must carefully choose
how to allocate their limited resources, and they tend to approve requests
that are in harmony with their own perspectives.

Based on the Kiersey-Bates analysis of temperaments, the odds would
favor the second memo (directed to a rational type), but writers need not
play the odds. If, instead, using the questioning methods outlined in Chapter 2, they make the effort to ascertain their readers' perspectives and write
accordingly, they will consistently get their requests approved.

8

ORAL PRESENTATIONS

Besides needing strong, versatile speaking skills in most day-to-day situations, law enforcement professionals need to be able to make effective oral presentations to the public. Career-day presentations, appearances before concerned citizens groups, talks to schools and PTAs, not to mention briefings to prosecutors and supervisors: all of these occasions call on officers and other personnel to speak well. In addition, more and more courses in all college curricula require oral presentations as colleges prepare their graduates for the frequent demand for public speaking in careers.

This writing guide includes a chapter on oral presentation because

1. Writing is an invaluable tool of preparation for public speaking.
2. Effective oral presentation often demands written materials, clearly and attractively designed, as part of the presentation.

This chapter will also briefly review some time-tested tips for effective speaking in public.

WRITTEN PREPARATION FOR THE TALK

For thousands of years, writing has been recognized as essential preparation for good speakers. The Roman rhetorician Quintilian (C.E. 35?–95?) regarded writing and speaking as inseparable:[*] "By writing we speak with

---

[*]*Institutio oratoria*, Book Ten; trans. H. E. Butler; (Cambridge, MA: Harvard University Press, 1921–1922).

greater accuracy, and by speaking we write with greater ease. We must write, therefore, as often as we have opportunity."

While technology has changed communication in many ways over the ensuing centuries, the necessity for professionals in any field to put across their ideas orally has not diminished. Indeed, in the age of television and the multimedia Internet, those abilities—along with writing skill—will be perhaps more stringently tested than ever before.

Writing to prepare for talk usually includes these tasks:

1. Detailed, focused *note taking and other predraft exercises*
2. Careful *outlining*
3. Either formal or informal *scripting* of the talk

Note Taking and Other Predraft Exercises

See Chapter 1, under "Writing and Memory: Taking Good Notes," for note-taking techniques. Also see Chapter 1 for detailed descriptions of other tools for data collection, such as annotation of reading and keeping a research log. See Chapter 2, under "Prewriting and Data Collection," for a description of other useful predraft techniques, such as the dummy draft.

If experienced writers typically devote 80 percent or more of their time on a project to predraft work, it's all the more important for speakers to invest this proportion of time. Speaking puts most of us under intense pressure to perform, and preparing thoroughly, so that we are confident of our materials, helps to relieve anxiety.

Writing to Anticipate Audience Questions

Certainly all writers need to think about questions that readers are likely to raise in response to the ideas that they write; such anticipation is a basic part of *audience awareness*, which we discuss from time to time in Chapters 1 through 7. But anticipating questions is especially important for the speaker, since an audience will demand an immediate response. Therefore:

- As you (1) collect data for a speech, as you (2) prepare a dummy draft or other predraft tool, and then as you (3) outline your talk, *keep a running list* of points that your audience is likely to find surprising or controversial—and that you may be asked to clarify or defend.
- Use this list to research and record information with which you might answer such questions. Even if you decide not to include this extra information in the talk itself, you will have done the search work necessary to back up controversial points.

Careful Outlining

Even if you use *visual aids*—charts, lists, pictures—to support your talk, listeners usually get only one brief chance to hear and remember what you've said. Moreover, since most people are less adept at recalling what they hear than what they see, your talk needs to be *outlined*—arranged—to give yourself the best chance to be memorable.

Two Outline Methods: Storytelling and Q and A

Outlining a talk in one of these two ways can help you prepare a more memorable presentation.

Telling Stories.   Everyone loves a well-told story; everyone hates a poorly told one, wherein the speaker leaves out important details or interrupts to backtrack: "Oh, I forgot to mention. . . ." Storytelling gives listeners a familiar framework on which to attach—and keep together—ideas. Arranging your talk as a kind of story helps readers to *remember* what you've said and *anticipate with interest* what you're about to say.

"How," you might ask, "can I make a story out of the report I have to give or the argument I have to make?"

Possibilities are everywhere. For example, the report of an experiment always contains the *procedures* used: "First, we combined the compounds in the test tube. . . . Second, we heated the mixture to 100 degrees C. . . . We ran into trouble when we added the . . ." and so on. A well-written play or movie review always includes enough of the plot to give the unfamiliar listener a sense of the action: "When Bottom appeared in the ass's head, the audience went wild." A field report in whatever discipline always includes a story element: "I saw the suspect use a key to unlock the 'Employees Only' door and enter the building at 7:01 p.m."

Outlining a talk as a story means taking advantage of those many story opportunities in any subject. For example, in arguing business points, effective speakers frequently use stories to *illustrate* their messages:

> Let's say that a customer complains about rude treatment by your service rep over the phone. What do you do? Now you could listen patiently, tell the customer that you'll look into it, hang up, and hope the customer cools off. But you could . . .

Using the story makes the problem real to the listener; it ensures greater attention than just stating the point: "See customer complaints not as a nuisance, but as an opportunity to strengthen relationships."

In a broader sense, however, outlining as a story means *presenting the entire report (or proposal) in a "narrative frame"—as a story.* You do this by making yourself a character, as it were, in the story. Note the difference between these two examples, drawn from a talk by a successful interrogator who was asked to share some secrets of success with other members of the department:

Version 1

The keys to successful interrogation are not overly complex, and although each case is unique, there are some generalizations about confessions that do apply. Hardly anybody confesses to an investigator who is uncertain of the suspect's guilt. Therefore, prior to any interrogation, investigators, using case facts and an interview of the suspect, must satisfy themselves of the suspect's guilt. Only when this has been done and the investigator has also taken the time to establish credibility with the suspect, can the interrogation begin.

An interrogation consists of making confession palatable to the suspect by downplaying the seriousness of the offense, by offering a good reason for the suspect's having committed it, or by placing the moral responsibility for the crime on someone or something else. When it becomes obvious that the suspect has accepted the investigator's explanation, the interrogator can then ask, "That is how/why it happened isn't it?" The answer requires only one word or even a nod, something much easier to do than to recount the details of some criminal act.

Version 2 (Story Frame)

I've learned that the keys to successful interrogation are not overly complex, and although each case I've been on has been unique, there are some generalizations about confessions that do apply. First, I hardly ever get confessions if I'm uncertain of the suspect's guilt. Therefore, prior to any interrogation, I use case facts and an interview of the suspect to satisfy myself of the suspect's guilt. Only when I've done this and when I've taken the time to establish credibility with the suspect can I begin the interrogation.

What's my goal in the interrogation? An interrogation consists of making confession palatable to the suspect. I do this in one or more of three ways: by downplaying the seriousness of the offense, by offering a good reason for the suspect's having committed the crime, or by placing the moral responsibility for the crime on someone or something else. When it becomes obvious to me that the suspect has accepted my explanation, I can then ask, "That is how/why it happened, isn't it?" The answer requires only one word or even a nod, something much easier to do than to recount the details of some criminal act.

Version 1 presents an outline for the talk that is clear: a sequence of logical steps. This outline is standard in written reports, where readers expect to find certain types of information in specific places. Readers can skim such reports because they know where to look for certain kinds of information; they can also go back and read more closely.

For a *talk*, however, the personal story outline is better suited to ensuring that the listener won't miss important details, because it involves the listener in the speaker's quest. Notice that the "story" may actually unfold very much as the standard report does: in this instance, the main difference between versions 1 and 2 is the shift to the first person.

What to Leave Out of the Story.   Nevertheless, because listeners can't be expected to hold onto nearly as much information as a written report would contain, the speaker will *leave out* from the talk many of the details, such as any pertinent studies by other researchers and most pertinent statistics. The speaker will rely more on *summaries* of research and findings and will back up the talk with *written material* (such as the written report itself) that listeners can read if they want to. As with any other good story, this one will be remembered by its hearers if it doesn't get bogged down in the mud of detail.

Q and A.   Question-and-answer structure is becoming ever more popular in business and academic writing and speaking. Like the story, it relies on age-old forms. In the case of Q and A, the structure depends on *dialogue*, our everyday mode of conversation. Indeed, the TV and radio talk shows have so institutionalized Q and A that in many contexts this format has replaced the lecture (or sermon or political speech) as the standard oral mode of delivering information and opinion.

Of course, a talk organized as Q and A merely creates the illusion of dialogue; but it's an effective ploy. Throwing out questions to your audience—especially if you include a "pregnant pause" before answering—in a way challenges your listener. It raises dramatic tension—and tension means attention.

Here's our excerpt from the interrogation talk as Q and A:

Version 3 in Q and A Format

How complicated is getting a confession from a suspect?	The fundamentals of interrogation are quite simple.
Where does one start?	To quote an old cliché, "At the beginning." By learning all one can about the case and the person.
Should one try to get a confession immediately?	No. The investigator should first interview the suspect for added insight and to give the suspect time to recognize the credibility of the investigator.

After doing this, should the interrogator ask the suspect if he committed the crime?	No. Instead, interrogators should offer plausible explanations for the crime that make it acceptable for the suspect to confess.
How do I know what is plausible?	By the information gathered during preparation and the interview. Based on the particular case, the investigator can choose to downplay the seriousness of the crime, give the suspect a good reason for having done it, or place the moral blame on someone or something else.
Will the suspect spontaneously confess?	No. When the interrogator senses that the suspect has accepted the justification offered, the interrogator can then ask, "That is why/how it happened, isn't it?"

The key to using Q and A well in your outline is to *organize the questions in a logical way.* Start by brainstorming pertinent questions as they come to you—don't worry about order—then rearrange the questions in a way that a curious listener might more logically ask them. This exercise may suggest further questions to insert in your emerging outline.

When you begin to fill in answers, you'll probably discover further questions; you may also find that you need to rearrange the questions once again; and so forth.

Three more advantages of Q and A:

1. Questions give you, the speaker, a great way to keep in mind all you have to say.
2. Questions make effective visuals on a screen or handout.
3. Questions give your listener a clear framework for organizing and remembering your talk.

SCRIPTING THE TALK

Speakers tend to work better from an outline than from a fully scripted text that they feel they need to memorize. Trained actors can bring off the spontaneity of a good speech through a memorized text, but few of us have the training or skill to make a set text come across to an audience as sincerely spoken, not to mention interesting. Too many fully crafted talks wind up being read to an audience—usually a deadly tedious experience—by a speaker who couldn't bear to leave out a word of the text but who didn't have the time or the skill to deliver it without reading.

On the other hand, if you are really comfortable speaking in public, and if you can make good eye contact (see the tips for effective speaking on pages 85–87) even while reading from a text, then a read, fully crafted speech may be most effective. In this case, "scripting" a talk indeed means creating, revising, and practicing a full script (see "Reading Your Talk" below).

Since most speakers are not proficient enough at delivery to read a speech effectively, *scripting* a talk means drafting and revising enough so that you feel you have discovered and organized everything you want to talk about. But it's a good idea to leave much of the actual wording unscripted—to allow yourself to "fill in" the gaps spontaneously as you are speaking, and therefore be able to maintain eye contact with your audience and make the speech seem spontaneous.

Here's an excerpt from a script for the interrogation talk (note that the Q and A framework has been used):

> Where does one start?
>> By learning as much as possible about *case* and *person:*
>>> Interviews
>>> Searches
>>> Background checks
>>> Other data in reports, etc.
> Should the investigator try for a confession right away?
>> No.
>> 1. First establish *your* clear sense of the suspect's guilt:
>>> Do a follow-up interview for added insight.
>> 2. Second, make sure suspect shows that he/she acknowledges your certainty and your credibility.

READING YOUR TALK

If you feel that you must read your talk—the problems with that option notwithstanding—then you may script the talk fully by revising your dummy draft. But always remember that the talking script is *not* a text that the audience can read. Use such formats as storytelling and Q and A to help ensure that your audience will listen attentively and remember your ideas.

Written Visual Aids

Although political speeches and religious sermons are still usually delivered without visual props, most professionals make their words more effective through graphics. These may run to the elaborate, such as videotapes, multilayered/multicolor graphs, Web pages on supersize projectors. Nevertheless, even simple uses of writing can add clarity and punch to a talk. For more complex uses of graphics, we've included a select list of sources at the end of this section.

If you want to keep it simple but still have impact, we suggest the following:

Your Simplified Outline

Don't include every subheading from your outline, but do provide your major topics (or questions, if you use Q and A) and some key terms you want your listeners to be sure to remember. A chalkboard, a poster, a flip chart, a transparency on an overhead projector, a one-page handout: any of these display options is suitable for a simple list. If possible, keep it to one page or screen that will remain in front of the audience throughout your talk and to which you can point as you move from idea to idea.

Interrogation Technique

Where do I begin?

Should I go for the confession right away?

Should I ask the suspect if he/she is guilty?

How do I know what is plausible?

Will the suspect spontaneously confess?

A *more detailed outline* may work well as a handout to accompany the talk. As long as your remarks follow the outline, listeners can use the handout for note taking.

Idea or Flow Chart

Use the two-dimensional space or the board, page, or screen to show relationships:

Study case and suspect in detail, **to:**
  Convince yourself of suspect's guilt **and**
  Prepare for interrogation; **THEN**

Interview suspect, **to:**
  Increase your insight **and**
  Convince suspect of your certainty **and** credibility

Creating the Visual as You Talk

A dynamic way to use visuals is to build the outline or flowchart as you talk. As you work from your own outline, fill it in for your audience visually as you come to each new topic or question. You can do this by writing on the chalkboard, flip chart, or overhead transparency so that by the end of your talk the audience has a complete outline.

Those adept at using overhead transparencies sometimes create a stack of these beforehand, with each new transparency adding another line or two to the outline. The speaker builds the stack, sheet by sheet, as the talk progresses.

Additional Information in Handouts

If you have more information available to your audience in written form, you can feel free to leave out many details from your talk that might intrude on their ability to remember your main points. Tell your audience about these additional materials wherever appropriate in the talk; however, we advise you *not* to distribute any of these materials during the talk, as doing so will distract the audience.

Tips for Effective Speaking

The following will not substitute for a fuller guide (see the list of sources at the end of the chapter), but they can serve as a partial checklist as you prepare and practice.

1. *Prepare, with writing as a help.* Using the tactics described up to this point is the best way to ensure a good talk.

2. *Speak distinctly, and more slowly than you would in normal conversation.* Out of nervousness, inexperienced speakers often hurry their talks and fail to enunciate their words. One great benefit of practice is consciously working on both pace and enunciation.

3. *Speak loudly enough for all to hear.* While it's not necessary to shout, it's important to project to all hearers. If you doubt your projection, ask, as you begin speaking, if those in the far reaches of the room can hear. Adjust as necessary. This procedure should also be followed if you are wearing a microphone or speaking into one.

4. *Maintain eye contact with all members of your audience.* Inexperienced speakers often let their eyes drift to one side of the room or the other; some even focus on one or two listeners and ignore the rest. Making *each and every person* in a room feel that you are talking to her or him takes conscious practice. One reason that we so strongly advise against reading a talk is that it eliminates the vital element of eye contact—unless the speaker is very experienced at maintaining eye contact while reading.

5. *Move as you speak.* Every speaking situation allows a different amount and kind of appropriate movement. A TV talk-show host routinely moves around a studio and from row to row among the audience; a graduation speaker stands behind a solid wood podium and speaks into a microphone; an attorney questioning a witness is free to move around the courtroom, but the witness must stay put; and so on. In no case should a speaker try to imitate a statue; nevertheless, most inexperienced speakers miss chances to keep their audience engaged through relevant movement.

Shifting eye contact and moderate use of hand gestures are the most common strategies that speakers use to keep their hearers' eyes attentive. But don't ignore the benefits of some moderate walking around a "speaking space" to keep your listeners engaged. Good teacher-lecturers routinely move from one side of a room to another, often writing down key ideas or pointing to terms already on a board or overhead. Business presenters do the same.

Even if you find yourself in a very rigidly defined space, or even if every preceding speaker has tried to be statuelike, don't hesitate to wake up your audience by doing something at least slightly different. Another benefit of having graphics tools such as a flip chart or overhead projector is that they give you a valid excuse for moving, even when others have been rooted to the spot.

6. *Remember to breathe.* This may seem a nonsensical suggestion, but the tense speaker often fails to breathe normally, with a consequent loss of composure, stamina, and volume. Speech trainers often use the formula "one thought, one breath": they urge their clients to take a breath at the end of each clause and sentence. Good breathing helps pacing, too.

7. *Warm up before "going on."* Like singers who loosen their vocal chords by doing exercises before performing, speakers should warm up their voices by practicing a small portion of the talk just before the official speech. This warmup can save you the embarrassment of a scratchy or "cracked" voice or of stumbling over syllables at the start of the talk.

8. *Observe other speakers.* In whatever situation you may be, look closely for techniques that make good talkers good. Whenever you find that a speaker is holding your attention, chances are that it's as much because of engaging technique as because of interesting content.

FURTHER SOURCES WE SUGGEST

Guides to Good Oral Presentations

Beebe, S., & Beebe, S. (1997). *Public speaking: An audience-centered approach.* Boston: Allyn and Bacon.

Detz, J. (1992). *How to write and give a speech: A practical guide for executives, PR people, managers, fund-raisers, politicians, educators, and anyone who has to make every word count.* New York: St. Martin's Press.

Gamble, T., & Gamble, M. (1994). *Public speaking in the age of diversity.* Boston: Allyn and Bacon.

Grice, G., & Skinner, J. F. (1995). *Mastering public speaking.* Boston: Allyn and Bacon.

Smith, T. C. (1991). *Making successful presentations: A self-teaching guide.* New York: John Wiley.

Sullivan, R. L., & Wircenski, J. L. (1996). *Technical presentation workbook: Winning strategies for effective public speaking.* New York: ASME Press.

Zeuschner, R. F. (1997). *Communicating today.* Boston: Allyn and Bacon.

Guides to Effective Design of Graphics to Aid Speaking

Beebe, S., & Beebe, S. (1997). Visual aids. In *Public speaking: An audience-centered approach* (pp. 305–327). Boston: Allyn and Bacon.

Briscoe, M. H. (1996). *Preparing scientific illustrations: A guide to better posters, presentations, and publications.* New York: Springer.

Robbins, J. (1997). *High-impact presentations: A multimedia approach.* New York: John Wiley.

Smith, T. C. (1991). Adding an extra dimension with visual aids. In *Making successful presentations: A self-teaching guide* (pp. 57–83). New York: John Wiley.

9

TAKING EXAMS

Law enforcement and criminal justice students are likely to encounter exams of different types in many of their courses. Some of these exams will require no writing, only marked responses to multiple-choice questions. Others may ask for short written answers. Still others may require essays of varying length and complexity. Some exams may combine these forms.

This chapter will deal with writing in each of these exam situations.

MULTIPLE-CHOICE EXAMS, TESTS, AND QUIZZES

Writing plays a role in one's preparation for such exams, even though the exam itself requires no writing. The procedures in Chapter 1 for systematic *note taking* and regular *summarizing* use the power of writing to help one understand, remember, and apply course material. This regular thinking-through-writing pays off in surer recall of information in timed testing situations.

Final Preparation Exercise

While the regular exercises described in Chapter 1 can make students feel more confident as they prepare for timed exams—thus reducing the last-minute anxiety common to test takers—writing can also serve a purpose in that final preparation. In reviewing *notes, summaries,* and other materials (textbooks, lab manuals, etc.), use *further note taking* as a means to remember and think about important details. Test takers often read aloud or attempt to visualize terms or explanations they worry about forgetting.

Instead, or in addition, take a few moments to write those items. Thus, you use the power of writing to more fully reinforce the knowledge.

WRITING SHORT, TIMED RESPONSES: THE PRO METHOD

Students may be called on to write brief *definitions* of terms or brief *summaries* of procedures, among other short writing tasks on timed tests. In all timed writing situations, the most essential tasks for the test taker are:

1. Understanding the question's content and *purpose*
2. *Recalling* the pertinent information to meet that purpose
3. *Organizing* the information to make a clear presentation

This *purpose/recall/organization* sequence—PRO for short—is a writing process specifically helpful in the timed situation. It works as follows.

Imagine the sample question:

> There is more to preparing for an interview than just <u>learning the facts</u> of the case. In a short <u>essay,</u> <u>identify</u> and <u>explain??</u> <u>three</u> <u>other</u> aspects of <u>preparing</u> for an interview.

1. Understand *purpose.*

Circle or underline each key term needed to fulfill the task successfully (*learning the facts, essay, identify, explain, three, other, preparing*). Identifying key terms helps you focus on all the important elements of the task.

Use a *question mark* to note any term that seems vague or difficult and that may cause you trouble in writing the response. For example, *explain* might mean "give a description of" or it might mean "justify the importance of." If the test-taking situation allows, you can use these markings to ask clarifying questions before attempting to write a response.

2. *Recall* all pertinent information.

Again, if the testing situation allows you to use scrap paper—or the reverse sides of the exam pages—for making notes during the exam, use that space for some predraft writing (see Chapter 2, under Predrafting) to *recall and apply* everything pertinent to answering the test question.

A good predraft technique to use here is the *dummy draft* (see Chapter 2). Write without editing; your goal here is to get as much information onto the page as you can in a short amount of time; this isn't the place to worry about organization or neatness or good spelling. If you aren't sure of the accuracy or pertinence of something you write in this draft, mark it with a question mark, but don't stop now to think it through; when you've drafted the answer to the entire question, then go back to your question marks and work through your doubts—*if* time allows.

Sample Dummy Draft of Short-Essay Answer

Three other aspects besides learning case facts. You need to be aware of putting the person at ease, so it's important to <u>learn facts about the person's background</u>—work, hobbies, family—so you can make the interview more of a conversation, not just questions. You also need to worry about <u>location</u>—the station might be better if you need to make them take you seriously, but someone else might freeze there and would be more open at home—it depends on the case and the person. Also, <u>timing</u> is crucial—too early and there aren't enough facts for a good interview, too late and the person might no longer want to talk openly, and you also need to worry about other distractions—you don't want to pressure a person when they have to get their kids at school, for example.

Remember that time is of the essence. You may only have time to jot down an informal outline (notice that the dummy draft here is little more than an outline, with phrases and clauses); nevertheless, be sure to devote a portion of your time on the question to some sort of predrafting.

3. *Organize* for a clear presentation.

After you've quickly written your dummy draft, go back to the question and double-check for clues about the *format* of the answer. In our sample, the format is straightforward. Here's a revised response to the assigned task, based on the dummy draft.

After learning the case facts, many other items should be considered when preparing for an interview.

Any preparation that helps interviewers <u>to sustain a conversation</u> rather than merely ask questions improves the chances of success. Developing background information regarding the person being interviewed will contribute to this possibility. This can include information about such things as work, family, and hobbies. Talking about sensitive topics with people we know is usually easier than discussing them with strangers. Interviewers must take the initiative in creating a feeling of familiarity. The more they know about the person being interviewed the easier this becomes.

<u>Location</u> can also play an important role in the outcome of an interview. Unfortunately, the best location varies from one case to the next

depending on the facts and the person. Does the interviewer want to impress the person with the seriousness of the matter? If so, the trappings of crime and punishment afforded by a police station might be best. On the other hand, if a feeling of security would seem to offer a better chance of success, then perhaps the person's home would be preferable. These are just two of the many possible locations for an interview. The interviewer must decide which location is best based on the facts rather than randomly selecting a location.

Just as in other endeavors, timing can be critical to an interview. If the interview is conducted too soon, essential facts might not have been developed. If done too late, a person's willingness to talk may have evaporated. Even the time of day can be critical. To expect a parent to give interviewers their undivided attention when they are due at school to pick up their children seems unrealistic.

The best organization might not always be so easy to predict by rereading the question. However, even when the writer is in doubt, a few design principles always apply:

1. You can safely give information *in the order in which it's asked.* Test readers usually must read quickly; if they don't find information where they expect to find it, they usually assume it's not present.

2. Lay out the writing *so that the reader won't miss each section of the answer.* If the question asks for "three other aspects" and an "explanation" of each, as in the sample just given, make sure your reader can easily see that you have included three other aspects in your answer. For example, start a new paragraph with each section *or* leave white space after each.

3. Even if the question offers no obvious format, *don't* put down information in the order in which you think of it; always arrange it in the order *in which a reader can most easily follow it.* Because writers usually think of ideas and details somewhat randomly and idiosyncratically, it's *essential* to take two steps: first, to recall, then to organize material. (See the next section, Writing Timed Essays, for a fuller discussion of options in organizing your answers.)

WRITING TIMED ESSAYS

Written examinations include timed and untimed (or "take-home") varieties. *Untimed exams* are similar to other types of paper assignments; writing them follows the writing process model described in Chapter 2. This chapter will not address them separately.

Writing *timed essay exams* also follows the PRO system:

1. Understand the *purpose* of the question or topic
2. *Recall* all pertinent information
3. *Organize* the information so the reader can follow it easily.

STEP 1: Understand Purpose (first fraction—
1–3 min.—of allotted time)

In their haste to start writing, many exam takers stumble at the first step: *understanding the question or topic through close, careful reading.* Some fail to address all parts of a question; others, not reading closely, assume a question that's not being asked. Whatever the failing, even strong essays by well-prepared students lose credit for "not addressing the question."

Method: Read the exam question with pen in hand. *Circle* or *underline* each key word as you read. Here's a sample:

Final Exam: The Literature of Crime

Essay 1 of 2 (33%)

Choose <u>one</u> of the <u>short story</u> authors we have read. Drawing evidence from <u>at least two</u> works by that author, <u>identify and discuss</u> ?? at least <u>five</u> (5) characteristic <u>detection methods</u> used by the author's <u>principal</u> detective.

Note the underlined words. The writer has made the effort to identify every element that must be considered in composing an effective essay.

Note that besides circling or underlining words, the writer has also used the question mark to identify key terms that seem vague to the writer. *If the rules of the exam allow, ask the exam giver to clarify terms that you have marked in this way.*

Toward understanding the *purpose* of the question, the most important words in any assignment are often the *action verbs.* In the sample question these verbs include *identify* and *discuss.* Other common action verbs that appear in assignments include *analyze, categorize* (or *classify*), *compare, define, describe, evaluate, explain, list, prove, recommend, reflect on, show* (or *demonstrate*), and *support* (or *substantiate*), among many others. Though you'll want to clarify these terms with the exam giver if they don't seem clear in the question, they mean roughly the following:

Analyze   Break a problem into appropriate parts toward reaching an overall judgment or solution.

Categorize (or **Classify**)   Put information into appropriate classes or categories, which the writer defines.

Compare   State the similarities *and* differences between two or more phenomena; often the writer is expected to take a position in favor of one of the phenomena being compared.

Define   In academic contexts, this does *not* mean looking up a term in *Webster's Dictionary,* though specialty dictionaries or encyclopedias in a discipline may be used as sources of definitions; "defining" usually requires that one not only report *what* the term means but also *why* it means that; a good definition usually shows how a term can mean different things in different contexts.

Demonstrate   See **Show.**

Describe   Report the stages of a process in detail or the physical features of an object, place, or person; precision and substantial detail are usually required.

Discuss   One of the most common key words in assignments and one of the trickiest to define, because testmakers use it to mean a range of actions (note that the writer in the sample marked it as ambiguous); it can mean *analyze, compare, describe,* or most of the other terms in this list; often it means that the writer should (1) *evaluate* the significance of research results and apply them to other contexts, or (2) *identify* the major differing positions on a controversial topic and eventually arrive at a judgment based on a careful comparison. *If at all possible, ask the exam giver to clarify this term.*

Evaluate   Give a judgment on the value of something, often in comparison with other examples or items. This judgment is not a simple "good" or "bad," but the result of reasoned analysis of pertinent factors—not only *that* an item is valuable, but also *how and why.*

Explain   Give reasons or causes for a phenomenon; often paired with *describe.*

Identify   Somewhat vague, it can be as formal as *define* or as informal as *list.*

List   Simply, to place in an order that is asked for in the question (e.g., "from most to least important"); often paired with *define* or *explain.*

Prove   Similar to *show* and *demonstrate;* often implied by *discuss.* In exam situations, it often means to state a hypothesis or take a position, and then give reasons or evidence to justify that hypothesis or position; in academic writing, *proof* often does not imply absolute or exclusive truth, but a reasonable judgment based on a careful, respectful weighing of different views.

Recommend   To state a preference among possible phenomena or courses of action; it usually implies that one give reasons and provide evidence to justify the recommendation.

Reflect on   Similar to *discuss;* equally vague. In some cases, *reflect on* means to consider, in writing, how a phenomenon might be significant or how it might be applied to other situations than the one in which it commonly occurs. Ask for clarification, if at all possible.

Show (or **Demonstrate**)   Depending on context, *show* means *prove, describe,* or *explain.* Read the question closely and ask for clarification, if necessary and possible.

Support (or **Substantiate**)   Similar to *explain;* often paired with *prove* or *take a position; support* means to give reasons or evidence to justify a position. See **Prove,** especially the definition of *proof.*

STEP 2: Recall All Pertinent Information (roughly one-third of allotted time)

Again, because of haste, many exam takers try to craft a "final" essay in response to a question without having gathered the information they need to write well. The PRO method includes a second step by which the writer calls up all the information pertinent to a strong answer, before he or she crafts the essay itself. This step saves writers time because it allows them to concentrate on information collecting rather than also having to worry about organization, tone, correct grammar, and other final draft elements.

Because time is of the essence and because most exam situations don't allow writers to consult source material, we will describe two writing methods to promote quick recall from memory in timed essay contexts:

Outline Listing

After having carefully read and marked the question, create a quick outline of the kinds of information you'll need to answer the question. A person answering the sample question might write:

Holmes

    Typical Methods

    "A Scandal in Bohemia"

      Methods Used

    "A Crooked Man"

      Methods Used

Then, in each category of the outline, brainstorm a list of the information one needs for a strong answer. *Remember: Brainstorming* means to write quickly without censoring what you write; your goal here is to get as much information onto the page in as short a time as possible. Here's a sample:

Holmes

Typical methods

careful record keeping (scrapbook), assiduous background checking, use of Watson as sounding board, confidante, agent; use of others as role players, informants; always working from clues, never
jumping to theory; always hypothesizing from clues and relying on
logic even when possible answers are far fetched.

"A Scandal in Bohemia"

Methods—letter, scrapbook, changing roles, preparation, using
precedent, "observing, not just seeing."

"The Crooked Man"

Methods—observing before theorizing, careful listening, careful
scrutiny of crime scene, respectful interviewing.

Mind Mapping (a.k.a. Clustering)

This well-known method allows the writer to locate information at various
spaces on the page rather than moving strictly top to bottom. The mind
map begins with placing the focal idea of the essay in the *center* of the page,
then placing subcategories and further details—as you think of them—
around the central idea:

```
                    background checks    prep of questions
"Scandal"                                              loyalty
changing roles                                                   skill
   groom                       set-up interviews       Watson
 clergyman            records                        streetboys
                    (scrapbook)            aid by others


              great eye        H's methods           evidence before
              for detail                                theory

                   close          logic          even if hypotheses
 "Crooked"        listener     relentless           far-fetched
 streetboys                                  "Scandal": failed burglaries
appeal to friendship                         "Crooked": missing room key
   love                                          strange stick
```

As you add information, you will become aware of connections among details and so will draw lines between terms to show the connection.

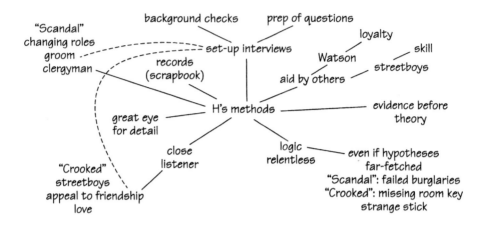

With these clusters of ideas displayed on the map, it is easy to turn them into an outline of principal topics:

H's Methods
Each method shown in "Scandal"
           in "Crooked"

There is no need to fill in the outline, since all the subtopics are already displayed on the map.

STEP 3: Drafting the Information in an Organized Format (roughly two-thirds of allotted time)

The first and second steps of the PRO method should comprise about a third of the time you set aside to devote to the exam question. The final two-thirds of the time should be given to *drafting* the essay itself, with a small portion of that two-thirds left for *editing* and *proofreading*.

Step 3 of the PRO method is very similar to the Drafting section of Chapter 2. It would be useful to review that section when preparing to take essay exams. The great difference between the process described there and the process for exam taking is that the exam situation usually allows no opportunity for seeking *feedback*, hence no meaningful opportunity for *revision*.

You must cultivate a strong sense of the demands of the *audience*—usually the professor—before you take the exam, because there will be little or no chance to refine this sense by asking questions during the writing period. Indeed, since most exams are taken well into and at the end of the semester course, one skill that is being tested is the student's understanding of what the professor expects.

Although the exam context doesn't allow the writer to fine-tune the essay for the professor, you can be sure of two essentials:

1. The essay should address *all* parts of the question.
2. It must be clearly organized.

We have addressed the first essential in Steps 1 and 2 of the PRO method. *Clear organization* is the province of Step 3.

Suggestions for Clear Organization

1. *Well-organized exam essays usually adhere closely to the order suggested by the question.* That's one reason why close reading and marking of questions (Step 1) are so important. In a different circumstance, the writer might want to construct and order the essay differently, but for the purposes of the timed writing situation, it's safest to put information where the reader expects to find it.
2. *Well-organized exam essays use paragraphs, white space, and other format features to show the structure clearly.* If you have carefully followed the Recall portion of the PRO method, then arranging information in a clear layout should not consume much time. Not only will the layout provide easier reading for the exam grader, it will also show that person that you have carefully considered your answer and understand how parts of the information relate logically.

Sample Exam Essay

A. Conan Doyle's detective Sherlock Holmes is best known for his ability to observe details that other investigators miss, and for his unparalleled powers to deduce truths—including solutions to great mysteries—from those details. While these talents consistently amaze the characters in the stories, they derive from methods consistently and relentlessly applied.

These methods include (1) careful and regular reading of various periodicals for "items of interest" as well as precise, well-organized categorizing of these items; (2) careful preparation for interviews by checking background and planning setting, attitude, and questioning approach; (3) assiduous listening and follow up questioning; (4) disciplined restraint from jumping to theories (i.e., only moving to theories when the evidence warrants); (5) practiced attention to physical details; (6) assiduous devotion to logic—even when a logical deduction seems far fetched; (7) changing of roles and tactics as the situation warrants; and (8) reliance on others as aids.

In two of the stories we read, "A Scandal in Bohemia" and "The Crooked Man," these methods show forth. In "Scandal," for example, Holmes prepares for his interview of "Count Von Kramm" (really the King of Bohemia) by reviewing his news files and using his Gazetteer to identify the source of the paper on which the mysterious letter had been written; thus, he already knows the identity of his disguised visitor and important background for the visit. His attention to detail is shown in his study of the letter, his external examination of the Adler house, and his use of the horse groom disguise to learn details about Adler's guests and habits. That he uses two very different disguises, the groom and the clergyman, shows his willingness to adapt his tactics to people and situations.

That he solves the location of the incriminating photograph is the result of all these methods; in addition, and perhaps most important, he uses logic, supported by precedent of other cases, to deduce that the photo is still in the house, even though two burglaries sponsored by the King had turned up nothing.

"A Crooked Man" also illustrates these methods, in greater or less proportion. Holmes enters into the mystery at Aldershot having studied the history of Colonel Barclay and his wife. Though he does not go to the elaborate lengths—e.g., disguises—that he does in "Scandal" to gather data and set up interviews, he does tailor his questioning of two key persons, Miss Morrison and Henry Wood, to what he knows or deduces of their backgrounds and interests. In both cases he appeals to their affection for Mrs. Barclay to get them to break silence—and he lets them speak without interrupting, and they do so at length once they are assured of Mrs. Barclay's danger.

That he solves this case is perhaps mostly the result of his letting the evidence lead to the theory. Whereas the police are all set to charge Mrs. Barclay with murder—despite the absence of the room key—Holmes will not overlook that detail nor the servants' claim that the strange walking stick had not been among the Colonel's possessions. Instead, these clues lead Holmes to speculate a third party, and so lead him to close inspection of the grounds and to questioning Miss Morrison—both actions essential to solving the case.

In both stories, there is another important method used that has not received as much acclaim as his relentless logic and powers of observation. Holmes relies on others, Watson principally, for loyal support. In "Scandal," not only Watson but a whole cast of "extras" play roles in the ruse to gain him admittance to the Adler house; in "Crooked," his always-handy group of street boys ensures that Wood won't go anywhere without Holmes' knowing. While Holmes the moody, often angry, usually solitary thinker has become legend, at least as important to solving crime is the Holmes who is able to win and maintain the loyalty of the brave, honest Watson and of a host of shadowy but indispensable aides.

Notice how the writer has used the data and patterns from Step 2 (Recall) of the PRO method to craft the essay. In addition to standard paragraph structure, the writer has also used numbering to highlight each method cited.

Notice also that in drafting the writer has gone beyond the details written in the outline and has given special emphasis to one method—Holmes's reliance on the loyalty and skills of others—by using it as the conclusion of the essay. This difference illustrates that the outline, or mind map, should not be used to restrict further thinking in the draft itself.

Editing and Proofreading

Most exam graders, realizing the pressure of time in exam situations, have a higher tolerance for minor errors in exams than they do in reports and other documents. Nevertheless, a few minutes reserved for proofreading can help the writer identify not-so-minor errors and make a stronger impression on the grader.

Several categories to observe:

- *Accurate use of technical terms:* e.g., procedures, diagnostic terms, chemical compounds, anatomy.
- *Accurate use of mathematical language:* Have statistics been cited accurately? Are numbers and symbols correct in formulae and equations?
- *Accurate references to authorities* (e.g., researchers) *and institutions* (e.g., agencies, boards).
- *Clear sentence structure:* e.g., have any words been inadvertently omitted? Are conjunctions, such as *however, although,* and *nevertheless,* used correctly, so that the reader can follow the writer's flow of thought? Are sentences long enough to convey the writer's sophisticated ideas, but not so long that they will confuse the reader?
- *Correct spelling:* This is of relatively low priority in timed exams, so infrequent spelling errors should not be a source of anxiety for test takers.

10

FINDING AND CITING
SOURCES FOR RESEARCH

This chapter has two parts. First, it provides a selected list of print (books, encyclopedias, periodicals, etc.) and electronic (Internet and Websites) sources useful to the researcher of topics in law enforcement. Second, it provides law enforcement examples for citing many different kinds of sources both in your written text and in an attached bibliography of "Works Cited." This chapter uses the citation rules of the American Psychological Association (APA), the style preferred in the social and behavioral sciences, including criminal justice and law enforcement.

LAW ENFORCEMENT/
CRIMINAL JUSTICE BIBLIOGRAPHY

Bibliography compiled by Brian Barker.

Books

Abadinsky, H., & Winfree, L. T., Jr. (1992). *Crime & justice: An introduction.* Chicago: Nelson-Hall.

Adler, F., Mueller, O. W. G., & Laufer, W. S. (1994). *Criminal justice.* New York: McGraw-Hill.

Cole, G. F. (1992). *The American system of criminal justice* (6th ed.). Pacific Grove, CA: Brooks/Cole.

Cole, G. F. (1996). *Criminal justice in America.* Belmont, CA: Wadsworth.

Cromwell, P. F., & Dunham, R. G. (1997). *Crime and justice: Present realities and future prospects.* Upper Saddle River, NJ: Prentice Hall.

Davis, J. A. (1995). *Conducting research: A preparation guide for writing and completing the research project or thesis in criminal justice, criminology, forensic science, and related fields.* Lido Beach, NY: Whittier.

Del Carmen, R. V. (1990). *Briefs of leading cases in law enforcement* (2d ed.). Cincinnati: Anderson.

Eskridge, C. (Ed.). (1996.) *Criminal justice: Concepts and issues: An anthology* (2d ed.). Los Angeles: Roxbury.

Eskridge, C. (Ed.). (1995). *Criminal procedure and practice* (3d ed.). Belmont, CA: Wadsworth.

Hancock, B. W., & Sharp, P. M. (1996). *Criminal justice in America: Theory, practice, and policy.* Upper Saddle River, NJ: Prentice Hall.

Harr, J. S., & Hess, K. M. (1990). *Criminal procedure.* St. Paul, MN: West.

Hess, J. E. (1997). *Interviewing and interrogation for law enforcement.* Cincinnati: Anderson.

Holtz, L. E. (1992). *Contemporary criminal procedure* (2d ed.) Binghamton, NY: Gould Publications.

Israel, J. H., Kamisar, Y., & Lafave, W. R. (1994). *Criminal procedure and the constitution.* St. Paul, MN: West.

Johnson, P. E. (1988). *Cases and materials on criminal procedure.* St. Paul, MN: West.

Kaci, J. H. (1994). *Criminal procedure: A case approach* (2d ed.). Placerville, CA: Copperhouse.

Kamisar, Y., Lafave, W. R., & Israel, J. H. (1994). *Basic criminal procedure* (8th ed.). St. Paul, MN: West.

Kamisar, Y., Lafave, W. R., & Israel, J. H. (1994). *Modern criminal procedure* (8th ed.). St. Paul, MN: West.

Katsh, M. E. (Ed.). (1995). *Taking sides: Clashing views on controversial legal issues* (7th ed.). Guilford, CT: Dushkin Publishing Group.

Keve, P. W. (1995). *Crime control and justice in America: Searching for facts and answers.* Chicago: American Library Association.

Klofas, J., & Stojkovic, S. (1995). *Crime and justice in the year 2010.* Belmont, CA: Wadsworth.

Meador, D. J. (1991). *American courts.* St. Paul, MN: West.

Neapolitan, J. L., & Schmalleger, F. (1997). *Cross-national crime: A research review and sourcebook.* Westport, CT: Greenwood Press.

Neubauer, D. W. (1988). *America's courts and the criminal judicial system* (3d ed.). Pacific Grove, CA: Brooks/Cole.

Samaha, J. (1990). *Criminal procedure.* St. Paul, MN: West.

Scheb, J. M., & Scheb, J. M., II. (1989). *Criminal law and procedure.* St. Paul, MN: West.

Schmalleger, F., & Armstrong, G. M. (1997). *Criminal justice today: An introductory text for the twenty-first century.* Upper Saddle River, NJ: Prentice Hall.

Senna, J., & Siegel, L. (1995). *Essentials of criminal justice.* St. Paul, MN: West.

Zalman, M., & Siegel, L. (1991). *Criminal procedure: Constitution and society.* St. Paul, MN: West.

Zalman, M., & Siegel, L. (1995). *Key codes and comments on criminal procedure.* St. Paul, MN: West.

Dictionaries and Thesauri

Black, H. C. (1990). *Black's law dictionary* (abridged 6th ed.). By the Publisher's Editorial Staff. St. Paul, MN: West.

Rush, G. E. (1994). *The dictionary of criminal justice* (4th ed.). Guilford, CT: Dushkin Publishing Group.

U.S. Department of Justice, Bureau of Justice Statistics. (1981). *Dictionary of criminal justice data terminology* (2d ed.). Washington, DC: Author.

U.S. Department of Justice, National Institute of Law Enforcement and Criminal Justice. (1978). *National criminal justice thesaurus.* Washington, DC: Author.

Walsh, D. (1983). *A dictionary of criminology.* London: Routledge.

Directories and Statistics

Hindelang, M. J. (1973–). *Sourcebook of criminal justice statistics.* Washington, DC: U.S. Department of Justice, Bureau of Statistics.

Lyles, S. (Ed.). (1985). *Directory of law enforcement and criminal justice associations and research centers.* Gaithersburg, MD: U.S. Department of Commerce, National Bureau of Standards.

U.S. Department of Justice, National Institute of Justice. (1993). *Criminal justice information exchange directory* (12th ed.). Rockville, MD: Author.

U.S. Department of Justice, Office of Justice Programs. (1994). *Directory of criminal justice information sources* (9th ed.). Washington, DC: Author.

Encyclopedias, Guides, and Almanacs

Bailey, W. G. (Ed.). (1989). *The encyclopedia of police science.* New York: Garland.

The guide to American law: Everyone's legal encyclopedia and supplement. (1983–1990). (Vols. 1–19). St. Paul, MN: West.

Kadish, S. H. (Ed.). (1983). *Encyclopedia of crime and justice* (Vols. 1–4). New York: Free Press.

Nash, J. R. (1990). *Encyclopedia of world crime: Criminal justice, criminology, and law enforcement* (Vols. 1–6). Wilmette, IL: Crime Books.

Schmalleger, F., & Armstrong, G. M. (Eds.). (1996). *Crime and the justice systems in America: An encyclopedia.* Westport, CT: Greenwood Press.

Literature Guides and Bibliographies

Lutker, M., & Ferall, E. (1986). *Criminal justice research in libraries: Strategies and resources.* New York: Greenwood.

O'Block, R. L. (1992). *Criminal justice research sources.* Cincinnati, OH: Anderson.

Vandiver, J. V. (1983). *Criminal investigation: A guide to techniques and solutions.* Metuchen, NJ: Scarecrow.

Schmalleger, F. (Comp.). (1991). *Criminal justice ethics: Annotated bibliography and guide to sources.* New York: Greenwood.

U.S. Department of Justice, National Criminal Justice Reference Service. (1975–). *National criminal justice reference service documents on crime and law enforcement.* Washington, DC: Author.

Periodicals and Journals

American Journal of Criminal Justice. (1984–). Richmond, KY: Southern Criminal Justice Association.
Crime and Delinquency. (1987–). Thousand Oaks, CA: Sage.
Crime Laboratory Digest. (1984–). Quantico, VA: U.S. Department of Justice, Federal Bureau of Investigation.
Crime, Law, and Social Change. (1991–). Boston: Kluwer.
Criminal Justice and Behavior. (Mar. 1974–). Thousand Oaks, CA: Sage.
Criminal Justice Ethics. (1982–). New York: John Jay College of Justice, Institute for Criminal Justice Ethics.
Criminology. (1970–). Columbus, OH: American Society of Criminology.
FBI Law Enforcement Bulletin. (1935–). Washington, DC: U.S. Department of Justice, Federal Bureau of Investigation.
Journal of Criminal Justice. (Mar. 1973–). New York: Pergamon.
Journal of Criminal Justice Education. (Spring 1990–). New York: John Jay College of Criminal Justice.
Journal of Criminal Law, Criminology, and Police Science. (1954–). Baltimore, MD: Williams & Wilkins.
The Journal of Research in Crime and Delinquency. (1964–). Beverly Hills, CA: Sage.
Justice Quarterly: JQ/Academy of Criminal Justice Sciences. (Mar. 1984–). Omaha, NE: The Academy.

World Wide Web Addresses

Bureau of Justice Statistics. <http://www.ojp.usdoj.gov/bjs/>
Copnet. <http://www.copnet.org>
Criminal Justice Institute. <http://www.cji.net>
Criminal Justice.Net. <http://www.webcom.com/cgm/>
FBI. <http://www.fbi.gov/homepage.htm>
Justice Information Center. <http://www.ncjrs.org/homepage.htm>
Law enforcement sites on the Web. <http://ih2000.net/ira/ira4.htm>
National Criminal Justice Commission. <http://www.ncianet.org/ncia/>
The Police Page: International Criminal Justice. <http://www.mcs.com/~jra/police/pages/main.html>
United Nations Crime and Justice Information Network. <http://www.ifs.univie.ac.at/~uncjin/uncjin.html>

CITING SOURCES USING APA (AMERICAN PSYCHOLOGICAL ASSOCIATION) STYLE

This section shows you how to cite a wide range of sources, print and electronic, using APA style, which is preferred in criminal justice and other social and behavioral sciences. The section shows how to cite sources both within your written text and in the attached "Works Cited" page that follows your paper or article. We have created examples using text from typical essays in law enforcement to show how to cite diverse kinds of sources.

More detailed information about APA style is available in *The Publication Manual of the American Psychological Association*, 4th edition (Washington, DC, 1994). The World Wide Web site for the APA publication manual is http://www.apa.org/books/pubman.html.

APA In-Text Citation Guide

The following examples show how to cite many kinds of sources within your written text. Whenever possible, sources should be cited within your text rather than in footnotes at the bottom of a page or in endnotes at the end of your paper or article. Citing sources in the text does not mean that you don't need to include a "Works Cited" page(s) at the end of your paper or article. Correctly citing sources in the text will help your reader locate those sources in the "Works Cited" page(s).

Citation guide written by Brian Barker.

A Source with One Author

In using the APA format for in-text citation, you must cite the author's name, the date of publication, and the page numbers for each source. Try to work most of this information directly into your text. The publication date, in parentheses, comes immediately after the author's name. Other parenthetical information should be cited immediately following the information it notes. When citing page numbers, use the abbreviation *p.* or *pp.*

> Shaw's study (1995) illustrates the relationship between alcohol abuse and domestic violence (pp. 41–49).
>
> A summary of previous research on hand gun control (Roberts, 1994, pp. 1–12) is evidence enough of their dangers.

A Source with Two Authors

In the parenthetical reference, use the ampersand (&) to join the authors names; in the text, however, spell *and* out.

> A new study predicts that gang violence will decrease in the next five years (Jones & Richards, 1997, p. 6).
>
> Jones and Richards (1997) concluded that gang violence will decrease in the next five years (p. 6).

A Source by Three to Five Authors

In the first reference, provide the names of all authors. However, in following references, you may use the lead author's name and the abbreviation *et al.* ("and others"). When citing a work with six or more authors, you may use *et al.* for all references.

> Hawkins, Lee, and O'Donnell (1987) did not examine the quality of psychological training new police recruits received. However, their survey does show that only 5 percent of the officers who quit within the first five years cited stress on the job as the primary reason (Hawkins et al., p. 56).

Two or More Works by the Same Author in the Same Year
In this case, list the sources in alphabetical order on your "Works Cited" page. When citing these sources in the text, label each source with a lowercase letter next to the year:

> Joe Smith's article "Are We Losing the Drug War?" would be cited using (1991a). In the same text, his article "Corruption within the DEA," would be cited using (1991b).

A Source by a Group or Corporate Author
In the first parenthetical reference, provide the organization's full name. Further, if the name can be abbreviated, place it alongside in brackets. In the following references, you may just use the abbreviation.

> In a recent newsletter, a 5 percent drop in crime was noted since the implementation of the community watch program (People Against Crime [PAC], 1997, 1).

Two Authors with the Same Last Name
In citing authors with the same last name, provide first and middle initials to distinguish them.

> The report credits the drop in the city's violent crime to the increased police presence after dark (J. Jenkins, 1993, p. 6). A new, nationwide study released recently echoes those findings (B. T. Jenkins, 1994, pp. 5–9).

Two or More Works in One Reference
In citing two or more works in one reference, use semicolons to separate the sources. The authors' names should be ordered alphabetically.

> In order to get citizens involved with fighting crime, police officers must build rapport with the members of the communities they patrol (Davis, 1987; Katz, 1989; Lawson, 1988).

A Multivolume Work
In the parenthetical reference, provide the volume number of the work from which you are quoting before the page number(s).

> Fonzelli states that Prohibition "watered the seeds of organized crime" (vol. 2, p. 57).

An Anonymous Work
When citing a work with no given author, use the first word of the title (excluding *a, an,* and *the*) in the parenthetical reference. In the following example the title of the source is *Your Community Safety Patrol.*

> The main objective of the Community Safety Patrol is "to assist the police officers in monitoring suspicious activities, and thus securing a safer environment within which to live" (*Your,* 1991, p.1).

Electronic Sources
Electronic sources and documents provide either page numbers or paragraph numbers, and in many cases neither. When page numbers are present, use the citation rules just outlined for texts. If the electronic source uses paragraph numbers, use the abbreviation par. or pars. and cite the paragraph(s) referenced.

> Last month, the *Law Enforcement Monthly* reported that three truck loads of illegal firearms were recovered in an undercover sting operation in Carter, Texas (Jones, May 1993, par. 6).

In cases where neither page numbers nor paragraph numbers are present, you can use the following abbreviations to indicate certain types of divisions within the source: *pt.* for part; *sec.* for section; *ch.* for chapter; and *vol.* for volume. In situations where page or paragraph numbers are not provided and no divisions are present within the source, reference the source with a name only.

> The committee concluded that the county's number one concern for the following five years is prison reform (Wiley, sec. 2).

No page or paragraph numbers are needed in citing an electronic source in which the entries are alphabetically ordered (i.e., online encyclopedia). If the item is unsigned, use the title or an abbreviated form of the title in your parenthetical reference.

> Penology, a branch of criminology, is concerned with prison management and the rehabilitation of criminals ("Penology").

APA *"Works Cited"* Format

Your "Works Cited" page(s) is attached to the end of your paper or article. It should include *all* sources that you have cited in the text of your paper or article and *only* the works you have cited. Don't include in your list any sources that you consulted in your research but did not cite in the text. Also, don't include in your "Works Cited" any material that you suggest that readers consult for further information, unless you have cited those works in the text itself. If appropriate to the assignment, you may add lists of "Additional Works Consulted" or "Suggested Additional Reading," but do not confuse these lists with "Works Cited."

When compiling your "Works Cited" page(s), cite sources alphabetically *by the last name of the author(s)*. (See Citation Examples for cases in which the author is unknown or there are multiple authors.) The first line of a citation always begins at the *left margin*. If the citation extends to more lines, indent subsequent lines *five spaces from the margin*.

The following examples were compiled by Brian Barker. For more detailed information on APA style (not specific to law enforcement), consult the *Publication Manual of the American Psychological Association*, 4th edition (Washington, DC, 1994). The APA manual Website is http://www.apa.org/books/pubman.html.

Books

By a Single Author

In referencing a book with one author, provide the author's name (last name first, first and middle initials only), the publication date in parentheses, and the title (underlined or italicized; capitalize the first word of the title and subtitle(s) only). Close the citation with the city of publication and the publisher.

Cole, G. F. (1996). *Criminal justice in America.* Belmont, CA: Wadsworth.

Two Authors

In citing a book with two authors, reverse both names. The names are separated by a comma and ampersand (&).

Abadinsky, H., & Winfree, L. T., Jr. (1992). *Crime and justice: An introduction.* Chicago: Nelson-Hall.

Three or More Authors

Name all authors, using the same format as that used for two authors.

Israel, J. H., Kamisar, Y., & Lafave, W. R. (1994). *Criminal procedure and the constitution.* St. Paul, MN: West.

Collection Produced by an Editor(s)
To distinguish the editor, place the abbreviation *Ed.* in parentheses immediately after the name.

Katsh, M. E. (Ed.). (1995). *Taking sides: Clashing views on controversial legal issues* (7th ed.). Guilford, CT: Dushkin Publishing Group.

Translated Work
In the case of a translated work, place the translator's name and the abbreviation *Trans.* in parentheses after the title.

Beccaria, C. (1963). *On crimes and punishments* (H. Paolucci, Trans.). Indianapolis: Bobbs-Merrill.

Collection Produced by a Corporate Author or Group
Begin the citation with the corporation or group authoring the work. In this example, the author and the publisher are the same; instead of repeating the name in the publisher's position, use the word *Author.*

U.S. Department of Justice, National Institute of Justice. (1993). *Criminal justice information exchange directory* (12th ed.). Rockville, MD: Author.

Multivolume Work
For a multivolume work, place the volumes in parentheses after the title.

Kadish, S. H. (Ed.). (1983). *Encyclopedia of crime and justice* (Vols. 1–4). New York: Free Press.

Book in a Series
In citing a book in a series, name the series and number in parentheses after the title.

Alpert, G. F. (1985). *The American system of criminal justice* (Law and Criminal Justice Series #1). Beverly Hills: Sage.

Work with Subsequent Editions
If the work has subsequent editions, indicate the edition number in parentheses after the title.

Del Carmen, R. V. (1990). *Briefs of leading cases in law enforcement* (2nd ed.). Cincinnati: Anderson.

Reprinted or Reissued Book
In citing a book that has been reissued, the original publication date should be presented first, followed by a slash mark and the new publication date.

Blanchard, R. E. (1974/1975). *Introduction to the administration of justice.* New York: John Wiley.

A Work from an Anthology or Edited Book
In citing a work from an anthology or edited book, the title of the selection should not be enclosed in quotation marks or underlined. Underline or italicize the title of the book. Following the title of the book, provide inclusive page numbers in parenthesis.

Kelman, M. (1982). Criminal law: The origins of crime and criminal violence. In David Kairys (Ed.), *The politics of law: A progressive critique* (pp. 174–212). New York: Pantheon Books.

Citing an Introduction, Preface, Foreword, or Afterword
In this case, follow the same format as for a selection in an edited book.

Bedau, H. A. (1982). Preface. In H. A. Bedau (Ed.), *The Death Penalty in America* (3d ed., v–viii). Oxford: Oxford University Press.

Reference Books and Encyclopedias
When citing an article from a reference book or encyclopedia, list the volume number and inclusive page numbers in parentheses following the title of the encyclopedia or book.

Teeters, N. K. (1992). Penology and prisons. In *Collier's encyclopedia* (Vol. 18, pp. 563–568). New York: MacMillan Educational Co.

Dissertation
Underline or italicize the title of the dissertation. Following the title, use the label *Unpublished doctoral dissertation* and provide the name of the university.

Wilf, S. R. (1995). *Imagining justice: Politics, storytelling, and criminal law in revolutionary America, 1763–1792.* Unpublished doctoral dissertation. Yale University.

Periodicals

Article in a Scholarly Journal, Continuous Pagination
Provide the author's last name and initials, the date of publication, and the title of the article(capitalize only proper nouns and the first word of the title and subtitle; do not enclose the title in quotation marks). Close the citation with the journal title and volume number (underline or italicize both) and inclusive page numbers.

Ford, J., Rompf, E., & Faragher, T. (1995). Case outcomes in domestic violence court: Influence of judges. *Psychological Reports, 77,* 587–594.

Article in a Scholarly Journal Paginated by Issue
In this case, you need to provide also the issue number of the journal. Place it in parentheses immediately after the volume number.

Ross, L. E. (1996). The relationship between religion, self-esteem, and delinquency. *Journal of Crime and Justice, 19*(2), 195–214.

Article in a Monthly Publication
In the following example, note that the year and month are inverted and the name of the month is spelled out. In citing page numbers for newspapers and magazines, use the abbreviations *p.* and *pp.*

Chiles, J. R. (1986, July). Anything can be counterfeited—And these days almost everything is. *Smithsonian*, pp. 34–43.

Articles in a Weekly Publication
Follow the same format for an article in a monthly magazine.

Shapiro, B. (1997, July 7). The adolescent lockup. *Nation, 265*(1), pp. 6–7.

Newspaper Article
Gonzalez, D. (1992, February 1). Acts of hate from the past: Bias crimes in New York City are nothing new. *New York Times*, p. A25.

Signed Editorial, Letter to the Editor, Review
In brackets, after the title of the article, provide the label *Letter* or the name of the book(s) being reviewed. If no title is given, the bracketed information follows the date.

McAllister, B. A. (1991, March 28). The chief cause of police brutality [Letter]. *Wall Street Journal*, p. A15.

Powers, R. G. (1990, May 6). More cops, more jails, more crime [Review of the books *The justice juggernaut: Fighting street crime, controlling citizens* and *The police mystique: An insider's look at cops, crime, and the criminal justice system*]. *New York Times Book Review*, p. 7.

Published Interview
Aynesworth, H., & Michaud, S. G. (1989, May). A killer's words [Excerpts from interviews with serial killer Ted Bundy]. *Vanity Fair*, pp. 146–147.

Article Abstract
At the end of the citation, in parentheses, provide information on where the abstract may be found.

Baer, B. F., & Klein-Saffran, J. (1990). Home confinement program: Keeping parole under lock and key. *Corrections Today 52*(1), 17–18, 30. (From *Criminal Justice Abstracts*, 1990, 22, Abstract No. 056-22).

Other Sources

If given, provide the name of the main contributor just as you would for an author, and in parentheses note the contributor's title (director, narrator, producer, etc.). After the title of the source, specify the medium in brackets (videotape, film, cassette recording, etc.); if a number necessary for retrieval is given, place the medium and number in parentheses. Close with the city and name of the distributor or publisher.

Television and Radio
A critical witness—The medical examiner's role [Television]. (1994, August 24). *Primetime Live*. New York: ABC.
ACLU challenges legality of teen curfews [Radio]. (1995, July 2). *Weekend Edition*. Washington, DC: National Public Radio.

Film, Video, or Sound Recording
Jersey, B. (Senior Producer), & McDaniel, J. (Narrator). (1997). *Crime and punishment in America* [Videotape]. New York: PBS Video.
Webb, J., Yarborough, B., & Burr, R. (Performers). (1975). *Dragnet: The story of your police force in action; Two complete radio dramas* [Record]. New York: Radiola.

Broadcast Interview
Koppel, Ted (Interviewer). (1994, November 18). A town meeting crime & punishment [Interview with inmates James F. Polk III, Lester Barnett, Jr. et al.]. *Nightline* [Television]. New York: ABC.

CD-ROMs

Citing a CD-ROM or Diskette
The CIA world factbook [CD-ROM]. (1992). Minneapolis: Quanta.

Citing an Article or Excerpt from a CD-ROM
Howell, V., & Carlton, B. (1993, August 29). Growing up tough: New generation fights for its life: Inner-city youths live by rule of vengeance. *Birmingham News* [CD-ROM], p. A1 (10pp.). New York: Silver Platter.

Online Sources: Internet

Provide the same information as you would for a printed source (or as much as available). Note that online journals, magazines, etc. sometimes use paragraph numbers instead of page numbers. Close with the date and address from which the information was retrieved.

World Wide Web, Article in a Reference Database

Chappell, D. (1998). Book of the year (1997): Law, crime, and law enforcement. *Britannica Online* [Online]. Retrieved January 12, 1998 from the World Wide Web: http://www.eb.com:180/cgi-bin/g?Doc=boy/97/K02900.html

World Wide Web, Article in a Journal

Boos, E. J. (1997, May 21). Moving in the direction of justice: College minds—criminal mentalities. *Journal of Criminal Justice and Popular Culture* [Online], 5(1), pp. 1-20. Retrieved January 12, 1998 from the World Wide Web: http://www.albany.edu/sci/jcjpc/vol5is1.html

World Wide Web, Article in a Magazine

Bocamazo, S. A. (1998, January 12). "Scientific evidence" is sharply limited by U. S. Supreme Court: "Junk science" under attack. *Lawyers Weekly USA* [Online], 48 paragraphs. Retrieved January 12, 1998 from the World Wide Web: http://www.lweekly.com/feature.htm

Information from a Computer Service

Perez-Pena, R. (1998, January 27). New York mayor and governor seek end to parole. *New York Times* [Online], p. B1. Retrieved January 30, 1998 from the *New York Times* database on AOL.

FTP (File Transfer Protocol) Sites

Certiorari denied: Drew, Robert Nelson v. Texas [Online]. (1994, August 1). [From the Supreme Court's Order List]. Retrieved January 29, 1998 from the World Wide Web: ftp://ftp.cwru.edu/U.S. Supreme.Court/ascii/080194.ZR.filt

Gopher Sites

Justice Ministry Spokeswoman. (1995, July 31). State attorney requests arrest warrant for Abu Marzook [Online]. [From Israeli Foreign Ministry Information Division]. Retrieved January 29, 1998 from the World Wide Web: gopher://israel-info.gov.il:70/00/constit/leg/950731.ter

Synchronous Communications

Boyd, B., & Winston, C. (1996, October 29). Rehabilitating sex offenders: Is it possible? *Justice* [Online synchronous debate], pp. 1–4. Retrieved April 14, 1998 from the World Wide Web: http://www.justice.com/debate/96/43/index0a.html

Discussion List Posting

Smith, R. (1995, April 1). Prison reform. *The International Electronic Criminal Justice Conference* [Online]. Retrieved May 29, 1996 from the World Wide Web: http://www.icjc.com/prison.txt

E-mail

Wise, M. (mwise@dmu.edu). (1998, January 29). Crime down on campus. E-mail to Joy Carr (jc2@perry.com).

INDEX